Tree Assessment Manual
Second Edition

Principal author
Julian A. Dunster

Contributing authors
E. Thomas Smiley
Nelda Matheny
Sharon Lilly

Tree Risk Assessment Manual
(Second Edition)

Development
Kathy Ashmore
Marni Basic
Aaron Bynum
Peggy Currid
Tricia Duzan
Amy Irle
Alex Julius
Wes Kocher
Amanda Padgitt
Luana Vargas

Design
Beatriz Pérez González

Illustrations
Bryan Kotwica

International Society of Arboriculture
P.O. Box 3129
Champaign, Illinois 61826-3129, USA
+1 217.355.9411
www.isa-arbor.com
permissions@isa-arbor.com

ISBN: 978-1-881956-99-0

©2017 International Society of Arboriculture. All rights reserved. Printed in the United States of America. Except as permitted under the United States Copyright Act of 1976, no part of this publication may be reproduced, distributed in any form or by any means, or entered in a database or retrieval system without the prior written consent of the International Society of Arboriculture.

Printed by Premier Print Group
10 9 8 7 6
1019/CA/3000

Preface

Trees are a significant aspect of life in urban and rural areas, providing many benefits to people and communities. Some of these benefits have direct monetary value—stormwater management, enhancement of air quality, climate moderation, and increased property values. Other benefits—such as aesthetics, sense of place, or spiritual significance—are less tangible.

Wherever trees are present, people, property, and activities are potentially at risk of injury, damage, or disruption. All trees have the potential to become an unacceptable risk at some point, but that does not mean we should automatically condemn them "to be on the safe side." Tree owners, or people acting on their behalf, have a duty of care to ensure that the trees in their care do not create an unreasonable risk. Tree owners, and society in general, must take a balanced and proportionate approach to managing and accepting risk.

It is not possible to eliminate all risks associated with trees. Therefore, the goal of tree risk management is to provide a systematic and defensible approach by which risks can be assessed and managed to a reasonable level. A sound understanding of risk and trees provides a robust foundation for achieving that goal. A risk assessor's assessments and recommendations must be systematic and structured—and based on good science and arboriculture—so that they can be defended in court, if necessary. Risk assessors should also familiarize themselves with all applicable local, regional, or national safety regulations that may affect worker safety.

Since its initial release in 2011, the qualitative methodology first outlined in ISA's *Best Management Practices: Tree Risk Assessment* has been widely adopted and is proving to be an effective approach to risk assessment within arboriculture, urban forestry, and vegetation management. Despite this success, ISA has made important updates and improvements to this version of the *Tree Risk Assessment Manual* while ensuring that it aligns with the revised *Best Management Practices*.

The following is a summary of what you can expect to find new or revised in this manual:

> **Guidance for applying Level 1 assessment.** This revision covers the option to hold factors constant during Level 1 limited visual risk assessments. Typically, one or two of the three factors assessed during a Level 1 tree risk assessment (likelihood of failure, likelihood of impact, consequences) is/are held constant. This technique allows the assessor to focus only on the factors that could differ among the trees assessed. For instance, along a right-of-way, the consequences of failure may be held constant, leaving the assessor with only the need to determine the likelihood of failure and likelihood of impact.
>
> **Emphasis on establishing a time frame.** Although the concept of time frame has always been an important element of tree risk assessment, it is emphasized more in this new version of the manual. The goal is for assessors to clearly state what time frame they are using for their risk assessment, specifically for the likelihood of failure.
>
> **Modified definition of *medium* likelihood of impact.** The definition for a *medium* likelihood of impact was changed from "the failed tree or branch may or may not impact the target, with nearly equal likelihood" to "the failed tree or branch could impact the target, but is not expected to do so." The change was made to adjust the scale of the likelihood of impact categories. The previous definition had set the bottom of the *medium* category at approximately 50 percent probability, which was too high. Moreover, based on advice from risk assessment professionals, the authors sought to avoid establishing quantitative thresholds, relying instead on verbal definitions and examples to help users calibrate the categories.
>
> **Assess trees considering "normal" weather.** This revision provides additional guidance to help tree risk assessors determine likelihood of failure, which is described in the context of weather events. Storms are classified into broad categories based on frequency of occurrence, wind speed, and precipitation—normal,

extreme, and abnormally extreme. Tree risk assessors are guided to assess likelihood of failure based on the normal weather that can be expected during the defined time frame. In some regions, that could include some extreme storms and, if so, those storms should be considered. Abnormally extreme storms, such as tornadoes, hurricanes, and heavy freezing rain, are not predictable and, in most cases, are not considered for categorizing likelihood of failure.

Emphasis on defining each risk assessed. Tree risk assessors may assess several different risks for an individual tree. Typically, multiple targets can be identified, and often there is more than one possible failure mode—that is, the location or manner in which failure could occur. It is important that assessors define what risk is being rated and to articulate those risks to the client in the report.

Tree risk assessment can be used for two reasons. On one hand, we want to be proactive by identifying high-risk situations before accidents occur and to recommend action so that damage to property or people can be avoided. On the other hand, we may want to assure the risk manager or tree owner that the mechanical integrity of the tree is acceptable (or can be made acceptable if some mitigation work is undertaken) within his or her threshold for acceptable risk. The benefits conferred by the tree are then retained without undue risk.

Risk assessment should not be used as an excuse to remove trees that are healthy and low risk. A tree owner may want a tree removed because it blocks a vista or drops debris, but the tree might be protected by local ordinance. Aesthetics and client desires should not bias your professional judgment.

The *Tree Risk Assessment Manual* is designed to provide tree risk assessors with the basic knowledge and methodology needed to perform a competent assessment. The information and methods covered are complex and have been synthesized from a wide array of technical literature. This systematic and structured approach to assessing tree risk is intended to help risk managers strike a balance between the risk a tree poses and the benefits that trees provide to individuals and communities. As knowledge of tree biology and risk management grows, the methodology will continue to serve the industry while evolving and adjusting to new research and ways of thinking about risk assessment.

Disclaimer

Practitioners of tree risk assessment do so at their own risk. Neither the International Society of Arboriculture nor the authors accept any responsibility, explicit or implied, for liability, loss, or consequential damage arising from the manner in which the materials presented in this manual are used in the field. The training materials provided are considered to be reasonable and sufficient to provide users with a thorough understanding of when and how to undertake tree risk assessments.

The contents of this manual are provided "as is" and "where is" and may change without notice as research and understanding of tree biology, tree mechanics, and risk assessment evolve.

Acknowledgments

Writing a manual of this complexity requires considerable input and review to ensure that the technical aspects are correct and that the writing is clear and concise. While there are, inevitably, differences of opinion about how some aspects are interpreted—and, in some cases, whether or not they should be included—we have made every effort to present the most current information available. Even so, we are very aware that as new research results and concepts emerge, there will be changes in the way that tree risk assessment is considered and practiced. Future editions of this text will incorporate these changes as they are adopted.

Frank Rinn, Jerry Bond, Terrence P. Flanagan, and Mark Hartley provided the primary technical review for this manual.

Additional technical review was provided by Norm Easey, Gerrit J. Keizer, Skip Kincaid, Christopher Luley, Ian McDermott, Dwayne Neustaeter, Doug Sharp, Dan Marion, Vince Urbina, and Luana Vargas.

Many experts from around the world provided review comments for the wood decay table in Appendix 2. Christopher Luley and Gerrit Keizer were extraordinarily helpful with their suggestions.

Table of Contents

Module 1. Introduction to Tree Risk Assessment 1

Module 2. Levels of Assessment 15

Module 3. Target Assessment 35

Module 4. Site Assessment 49

Module 5. Tree Biology and Mechanics 65

Module 6. Tree Inspection and Assessment 99

Module 7. Data Analysis and Risk Categorization 121

Module 8. Mitigation 139

Module 9. Reporting 155

Appendix 1. Using the ISA Basic Tree Risk Assessment Form 163

Appendix 2. Common Wood Decay Fungi 177

Glossary 179

Selected References 187

Index 189

Introduction to Tree Risk Assessment

– *Module 1* –

Introduction to Tree Risk Assessment

Module 1

Learning Objectives

- Describe how tree risk assessment fits into the larger context of tree risk management, and explain how the roles of the risk manager and the risk assessor differ.
- Define the basic terms used in tree risk assessment, and explain the difference between "risk" and "hazard."
- Discuss the elements that should be included in the scope of work, including why each should be defined and agreed upon in advance.
- Compare and contrast quantitative and qualitative risk assessment approaches.
- Describe the steps involved in the risk assessment process.
- Explain the professional and legal responsibilities related to tree risk assessment.
- Discuss the potential safety issues involved with tree risk assessment.

Key Terms

acceptable threshold	hazard	qualitative risk assessment	target
breach of duty	inspection	quantitative risk assessment	time frame
consequences	legal precedents		tree conflict
duty of care	liability	risk	tree risk assessment
ethics	likelihood	scope of work	tree risk evaluation
failure	limitations	standard of care	tree risk management
harm	negligence		

Introduction

The benefits that trees provide to people living and working in the urban environment range from ecological and monetary to aesthetic and sociological. These benefits increase as the age and size of the trees increase. However, as trees get older and larger, they may become more likely to shed branches or develop decay or other conditions that can increase the likelihood for failure. The consequences of large tree or branch failures can also be greater.

It is impossible to maintain trees free of risk; some level of risk must be accepted to experience the benefits that trees provide. The National Tree Safety Group (NTSG), which is a partnership of organizations in the United Kingdom, has drafted a guidance document (2011) that identifies five key principles for tree risk management. This document provides a foundation for balancing tree risks and benefits:

- Trees provide a wide variety of benefits to society.
- Trees are living organisms and naturally lose branches or fall.
- The risk to human safety is extremely low.
- Tree owners have a legal duty of care.
- Tree owners should take a balanced and proportionate approach to tree safety management.

Tree owners/managers should strive to strike a balance between the risk that a tree poses and the benefits that individuals and communities derive from trees.

Tree Risk Assessment within the Context of Tree Risk Management

Tree risk management is the application of policies, procedures, and practices used to identify, evaluate, mitigate, monitor, and communicate tree risk (Figure 1.1). Tree risk assessment is the systematic process used to identify, analyze, and evaluate tree risk; it is a subset of tree risk management. Various people share responsibilities for tree risk management—including the tree owner or manager, the tree risk assessor, and the arborist providing service work (Table 1.1). Each plays a role to help assess, manage, and mitigate tree risk.

Many trees are located where the potential consequences of failure are *minor* or *negligible*. But in urban and developed areas where people could be injured, property damaged, and activities disrupted, the consequences of **tree conflict** (interference with people, property, or activities) or failure may be *significant* or even *severe*. Decisions on whether a tree **inspection** is required or what level of assessment is appropriate should be made with consideration for what is reasonable and proportionate to the specific conditions and situations. These are tree risk management issues.

Figure 1.1 In assessing and managing trees, you should strive to strike a balance between the risk that a tree poses and the benefits that individuals and communities derive from trees.

Table 1.1 Guidance on the intended roles of the tree risk manager, tree risk assessor, and the arborist. In addition, legal counsel can provide advice on duty of care, professional responsibilities, negligence, title and boundary matters, and other issues. All recommendations should be made in accordance with applicable industry standards and regulations.

Tree Risk Manager (tree owner, property manager, controlling authority)	Tree Risk Assessor (unless regulated by controlling authorities)	Arborist/Tree Worker
Duty of care responsibility Define and communicate tree risk management policies Determine the need to inspect the trees in question Establish the budget Identify the geographical limits of the tree inspection Specify the desired level of assessment Determine or accept the scope of work (shared with risk assessor) Decide the level of acceptable risk Establish the inspection frequency Verify target zone uses and occupancy rates Prioritize work Choose among risk mitigation options	Develop or accept scope of work, including time frame (shared with risk manager) Identify tree and site conditions to inspect Identify significant targets, estimate occupancy rates and target zone Assess and classify the likelihood of a tree failure impacting a target Assess the potential consequences of a tree failure impacting a target Analyze tree risk Consider if advanced assessments are needed Develop options for treatments to mitigate risk Estimate residual risks after treatment Recommend an inspection frequency Develop report Send report to client and explain findings to the client, if needed	Provide requested services: • Tree work safety review • Pruning • Removal • Support systems • Lightning protection • Tree health treatments • Transplanting • Tree replacement Identify the need for follow-up treatments

Fortunately, serious damage, injury, or death from tree failure is relatively uncommon. Tree failures during "normal" weather conditions (including typical storms for the area) are sometimes predictable and preventable due to obvious defects or conditions that can lead to failure. However, any tree, whether it has visible weaknesses or not, will fail if the forces applied exceed the strength of the tree or its parts. For example, gale-force winds, heavy snowfall, or freezing rain can break even solid, defect-free trees.

Tree risk assessments are usually undertaken because the owner/manager recognizes that the trees have potential to cause damage and/or injury. In some cases, to assess the potential for additional problems and manage the risk, assessments may be undertaken after one or more trees have failed. Anticipating problems, identifying specific trees where risk exceeds the **acceptable threshold**, and preventing damage before failure occurs are the key reasons for undertaking assessments.

Basic Terms Used in Tree Risk Assessment

To understand the fundamental tree risk assessment concepts, you must first know how the terms used in tree risk assessment are defined. Following is a partial list of these terms and their definitions. A more comprehensive list can be found in this book's glossary.

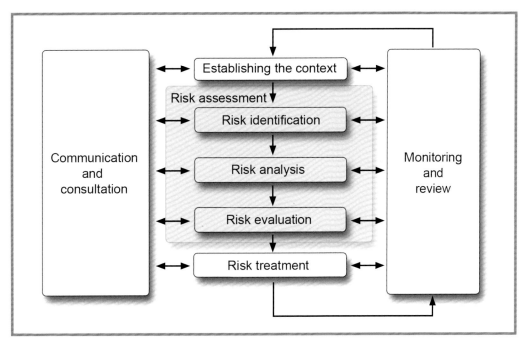

Figure 1.2 Contribution of risk assessment to the risk management process. ©IEC. This material is reproduced from IEC 31010:2009 with permission of the American National Standards Institute (ANSI) on behalf of the International Electrotechnical Commission. All rights reserved.

- **Risk** is the combination of the likelihood of an event and the severity of the potential consequences. In the context of trees, risk combines the likelihood of a conflict or tree failure occurring and affecting a target with the severity of the associated consequences—personal injury, property damage, or disruption of activities.

- **Tree risk assessment** is the systematic process used to identify, analyze, and evaluate tree risk. Risk is assessed by categorizing or quantifying both the *likelihood* of occurrence and the *severity* of the consequences.

- **Tree risk evaluation** is the process of comparing the assessed risk against given risk criteria to determine the significance of the risk. The magnitude of risk can be categorized or calculated and compared to the client's tolerances (or other predefined tolerances or ordinance requirements) to determine if the risk is acceptable.

- **Tree risk management** is the application of policies, procedures, and practices to identify, evaluate, mitigate, monitor, and communicate tree risk.

Figure 1.3 Decisions on whether tree inspections are required, or what level of assessment is appropriate, should be made with consideration for what is reasonable and proportionate to the specific conditions and situations. Decisions must also be made within the framework of applicable laws, standards, and regulations.

- **Targets (risk targets)** are people, property, or activities that could be injured, damaged, or disrupted by a tree failure.

- **Failure (tree failure)** is the breakage of stem, branches, or roots, or loss of mechanical support in the root system.

- **Harm** is personal injury or death, property damage, and/or disruption of activities.

- **Likelihood** is the chance of an event occurring. In the context of tree failures, likelihood refers to (1) the chance of a tree or tree part failure, (2) the chance of that tree or tree part impacting a specific target, and (3) the combined likelihood of a tree failing and the likelihood of impacting a specific target.

- **Consequences** are the effects or outcome of an event. In tree risk assessment, consequences include personal injury, property damage, or disruption of activities due to the event.

Approaches to Risk Assessment

Before a tree risk assessment takes place, it is important to establish the context of the assignment. Context defines the parameters of the risk assessment, including objectives, how risk will be evaluated, communication flow, applicable policies or legal requirements, and limitations of the risk assessment. The manner in which the risk assessment process is applied depends on the context and the methods used to carry out the risk assessment.

The two primary approaches to risk assessment are quantitative and qualitative. Each has advantages and limitations, and each may be appropriate with different objectives, requirements, resources, and uncertainties. Both the quantitative and qualitative approaches are valid when applied properly with reliable data and valid assumptions. Training and experience can improve the reliability of each approach.

Quantitative risk assessment estimates numeric values for the probability and consequences of events, and then produces a numeric value for the level of risk, typically using the following formula:

$$\text{Risk} = \text{Probability} \times \text{Consequences}$$

An advantage of quantitative assessment is that tree risk can be compared not only to other trees but also to other types of risk, as might be necessary for municipal decisions in which resources must be allocated among departments, for example. The calculations can vary from simple to complex because risks are analyzed independently or in combination.

Even if complex statistical analyses are carried out, risk assessors must remember that the calculations are estimates, and their accuracy and precision may not be consistent with the data and methods employed. Our ability to quantify probability is often limited when applied to trees because they are natural structures, and we have little systematically collected data on which to base probabilities. Because numeric data are not always available, and both systematic and statistical uncertainties can be high, full quantitative analysis is often not warranted or practical for tree risk assessment.

Qualitative risk assessment is the process of using categorized ratings of the likelihood and consequences of an event to determine a risk level and to evaluate the level of risk against qualitative

> ### Risk vs. Hazard
>
> In the past, arborists and foresters have used the term *hazard tree assessment* to describe the process of inspecting and evaluating the structural condition of a tree and the harm that could occur if it failed. The more accurate and appropriate term *risk assessment* is now standard.
>
> Risk is the combination of the likelihood of an event and the severity of the potential consequences.
>
> A **hazard** is a likely source of harm. In relation to trees, a hazard is the tree part(s) that might fail, and which is identified as a likely source of harm.
>
> A tree is considered hazardous when it has been assessed and found to be likely to fail and cause an unacceptable degree of injury, damage, or disruption—that is, it poses a high or extreme risk.

criteria. The term *likelihood* is used rather than *probability* because probability may imply quantitative odds. Often, ratings are combined in a matrix to categorize risk. This approach is a recognized and respected method of risk assessment used internationally by many governments and businesses.

It should be recognized that inherent subjectivity and ambiguity are **limitations** of the qualitative approach. To increase reliability and consistency of application, it is important to provide clear explanations of the terminology and significance of the ratings defined for likelihood, consequences, and risk.

Several of the qualitative numerical tree risk assessment systems previously used throughout the world assigned numbers to certain factors to derive an estimate or ranking of relative risk. The rankings were sometimes used to prioritize work. The assigned numbers, which are actually categorizations and do not represent a mathematical relationship, are either added or multiplied to develop an overall relative level of risk. Risk professionals caution that addition or multiplication of these ordinal numbers (rankings) is mathematically incorrect. Some of these systems were designed to estimate level of risk for individual trees and others to prioritize work within a population of trees. If a qualitative numeric system is employed, it should be used only for the intended purpose and with an understanding of its limitations.

The methodology of the risk assessment selected should be appropriate to the situation and should consider the goals and the resources available. With the context defined, the specific techniques should be selected based upon the following:

- Needs of the decision makers and the level of detail required
- Resources available and what is reasonable for the potential outcomes
- Availability of information and data
- Expertise required
- Applicable laws or regulations

A considerable level of uncertainty is typically associated with tree risk assessment due to our limited ability to predict natural processes (rate of progression of decay, response growth, etc.), weather events, traffic and occupancy rates, and potential consequences of tree failure. Sources of uncertainty should be understood and communicated to the risk manager/tree owner.

A matrix-based, qualitative approach to tree risk assessment has been selected for expanded explanation in this manual, but other approaches are not precluded from best management practices. Whichever technique is chosen, assessors should recognize the limitations as well as the nature and degree of uncertainty in the data and information available.

In practice, risk assessment always involves judgment, often using data that are not as good as we might like. All estimates of risk must be based on the best available data, and it should be recognized that estimates have some inherent uncertainty. They are always based on best professional judgment.

A primary goal of tree risk assessment is to provide information about the level of risk posed by a tree over a specific time frame. This is accomplished in qualitative tree risk assessment by first determining the likelihood of tree failure and impact, and then combining that likelihood with the consequences of tree failure. These assessment factors are determined by:

1. Evaluating the structural conditions that may lead to failure; the potential loads on the tree; and the tree's adaptations to weaknesses—to determine the likelihood of failure. Guidance for evaluating tree condition and assessing likelihood of failure are presented in Modules 5 and 6.

2. Evaluating the likelihood that a failed tree or tree part could strike people or property or disrupt activities.

3. Assessing and categorizing the value of potential damage or harm to the targets to estimate the consequences of failure.

With ratings for the likelihood of failure and impact

and consequences of failure, the level of risk then can be estimated or categorized.

It is important to define what risk is being rated. For example, "The risk of the large, dead branch in the oak tree in Mrs. Smith's back yard failing and hitting the garage" may be very different from "The risk of the oak tree in Mrs. Smith's back yard failing at the base and hitting people using the patio below." Multiple risks may be rated for any given tree, but the assessment report should state clearly what risk(s) have been rated to avoid ambiguity or confusion.

Types of Risk Associated with Trees

Trees can pose a variety of risks, which are categorized into two basic groups: conflicts and structural failures. The majority of this publication focuses on structural failures of trees.

Conflicts

Risk can arise when conflicts develop between trees and societal functions. As trees grow, they may produce potentially problematic flowers, fruit, roots, branches, and leaves, and these may conflict with pavement, the structures around them, and the people who use them. Conflict can also occur between power lines and trees. This contact can interrupt power supplies, can produce fires, and can injure or kill people. In some situations, these conflicts may cause other problems that create additional risks. Risk associated with conflicts is not the focus of this manual, though the

Figure 1.5 Flowers, fruit, and seeds from trees can cause slipping hazards and other conflicts with people in the community.

fundamental steps and methods of assessment can be applied to such assessments.

Failures

Structural failures occur when the stresses due to the forces acting on a tree exceed the strength of the tree structure or the tree–soil connection supporting the tree. Even a structurally strong tree that is free of defects will fail when a load that exceeds the load-carrying capacity of one or more of its parts is applied.

Most tree structural failures involve a combination of structural defects or conditions (such as the presence of decay or poor structure) and an unusual or extreme loading event (such as a strong wind).

Defining the Scope of Work

Prior to beginning a tree risk assessment, it is important that you work with the tree owner/manager (client) to define the **scope of work**. You must agree on the goals, limitations, and budget of the tree risk assessment. Some clients may be very experienced and will have already developed a sophisticated scope of work. Often, however, the risk assessor plays a significant role in documenting the scope of work.

Figure 1.4 Heaving of sidewalks can create a risk of tripping.

Regardless of who drafts it, be sure to review and accept the scope of work before beginning work.

Any property boundaries that restrict access to the tree(s) should be identified. Consider requirements for inspection, reporting, and permitting from the local government or authority. If a written report is to be presented to someone other than the person with whom you contract, you should identify that person or agency.

Your scope of work agreement should specify the risk assessment level(s) to be conducted. The level(s) should be appropriate for the assignment. Three levels of tree risk assessment are defined in ISA's *Best Management Practices: Tree Risk Assessment*, and these levels are described in more detail in Module 2 of this manual.

- Level 1: Limited visual assessment
- Level 2: Basic assessment
- Level 3: Advanced assessment

In addition to specifying the level of assessment, you also should describe pertinent details regarding the methods. For example, a Level 1 assessment can be done by walking by, driving past, or flying above the trees. The method used will greatly influence the cost and reliability of the results.

The scope of work should include specifications for the following:

1. **Identification of the tree(s) or area to be assessed**. This may be the location of a tree (for example, "the large oak tree in the front yard"), or it may include selection criteria (for example, "all trees on Main Street greater than 12 inches [30 cm] in diameter"). When assessing trees for a municipality or large property, it is important to have maps with clearly defined boundaries and clear directions for assessing boundary trees.

2. **Level and details of the assessment**. One or more of the three levels of assessment should be specified, as well as details that are to be included within each level. If the lowest level of inspection (limited visual) is selected, how the inspection is to be done and what information is to be recorded should be described. For example, "The inspection procedure is a walk-by from the sidewalk, looking for any obvious, aboveground defects."

Figure 1.6 Most tree structural failures involve a combination of structural defects or conditions, such as the presence of decay or poor structure, and an unusual or extreme loading event, such as a strong wind.

Figure 1.7 Although much attention is focused on whole-tree failures, the majority of failures are branches.

Details of conditions to be assessed also may be included in the scope of work. For example, trees often have many small, dead branches present, but you will likely be concerned only about larger branches that can result in serious consequences if they fail. The assignment may include limits on what is to be assessed; for example, "Only branches greater than 2 inches (5 cm) in diameter should be noted." Similar levels of detail can be specified for other conditions of concern.

For all levels of assessment, if you determine that a higher level of assessment or different type of assessment is needed, you should make that recommendation to the client.

3. **Method of reporting**. The manner of reporting and any additional documentation should be defined. The preferred method is a written report. In some instances, however, the report may be verbal (oral) with a recommendation for mitigation, or a work order for the mitigation. In general, verbal reports are not recommended because of the potential for misinterpretation in the chain of communication.

4. **Timetable for inspection and reporting**. The time frame for the inspection and delivery date for the report should be specified. For instance, the agreement might state, "Work to be done before trees leaf out, with report to be submitted within two weeks of completion of field work."

5. **Risk rating and mitigation**. Risk assessments typically include a rating of the current tree risk, options and/or recommendations for mitigating risk, assessment of the residual risk after mitigation, and the recommended inspection interval, if applicable.

Establishing the Time Frame

It is essential to establish a **time frame** for the risk assessment. The time frame provides a point of reference for the level(s) of risk determination. All trees could fail eventually, so you must associate your risk ratings with a time frame. For example, "This risk assessment covers a three-year period and is based on conditions present at the time of assessment." At times, the client may state the time frame in the specifications, especially for large-scale populations of trees. More often, you will either agree upon a time frame with the client, or you will establish the time frame.

As an additional service to a client, you may offer to rate the risk for more than one time frame, which may be helpful in making mitigation decisions. For example, the risk could be rated for a one-year period and a three-year period if it makes a significant difference. This may be more practical for risk assessments of a single tree or small number of trees. Note, however, that as the time frame increases, so does the uncertainty. There may also be reason to vary the time frame among trees within an assessment assignment, although this is not typical.

Be careful about defaulting to a one-year period for all assessments. While this may be appropriate for some specific trees, it is unrealistic and unnecessary for all trees, and it may place the client in an untenable position. Legal regulations or standards may define inspection intervals.

The time frame should not be considered a "guarantee period" for the risk assessment.

The Tree Risk Assessment Process

A primary goal of tree risk assessment is to provide information about the level of risk posed by a tree over a specific time frame. This is accomplished in qualitative tree risk assessment by first determining the likelihood of failure and impact, and then estimating the consequences of tree failure. These factors are used to categorize the level of risk. An overview of the process can be summarized as follows:

1. Identification of the potential targets.

2. Assessment of the site for factors that could contribute to or mitigate risk.

3. Evaluation of the structural or site conditions that may lead to failure, the potential loads on the tree, and the tree's adaptations to weaknesses—to categorize the likelihood of failure.

4. Assessment of the likelihood that a tree or tree part could strike people or property or disrupt activities to categorize likelihood of impact.

5. Evaluation of the targets' values and potential damage to categorize the consequences of failure.

6. Evaluation of the assessed risks in comparison to the client's risk tolerance.

7. Reporting findings, including recommended mitigation options and their associated residual risks.

Details about this process are provided in succeeding modules in this manual.

Professional Responsibilities

When you undertake risk assessment as a professional practice, you have responsibility for your opinions and recommendations. Sometimes, there can be legal implications to your work. The assessment reports will be used as the basis for decisions regarding public safety, as well as the safety of buildings and other structures. Your work may be reviewed by insurance professionals, lawyers, and (potentially) the courts. By signing and submitting your assessment report, in whatever form that takes, you are explicitly stating that you have undertaken a thorough and technically valid assessment within the scope and level defined. In the event of a claim, it is important to be able to show that a professional assessment has been undertaken.

Maintaining professional **ethics** and integrity is essential. Your assessments should be impartial and objective and your conclusions based on the relevant facts, observations, and analyses. You have a duty to maintain your independence as a tree risk assessor and to avoid conflicts of interest. If you or your employer provides other arboricultural services that might be recommended as mitigation measures, you should disclose this information to the client and ensure that your assessment and recommendations are independent of any other potential service.

There may be times when you recognize that you do not have the appropriate skills or equipment to undertake some aspects of the risk assessment.

Examples include conducting a heartwood decay assessment, climbing a tree for aerial inspection, or performing a detailed root examination. In such situations, you should reserve judgment and, if approved by your client, bring in a specialist who can undertake those aspects for you.

Integral to managing risk associated with trees is managing **liability** should a tree fail. Many countries and some jurisdictions in the United States have passed legislation that addresses who is responsible for maintaining trees. In other places, courts base their rulings on **legal precedents**—similar cases that have been decided in the past. You should be aware of the applicable laws in your country, state, and local jurisdiction.

When evaluating a claim or suit, courts consider the duty of care and the standard of care. **Duty of care** for a tree owner means that the owner has some level of responsibility to ensure a reasonable degree of safety for people or property near the trees under his or her care. The duty of care for a tree risk assessor is to use the generally accepted standard of care (as defined in applicable standards, best management practices, qualifications, and training courses) when performing work. Whether or not the duty of care has been properly met is judged by examining if and how the standard of care was implemented. Failure to act reasonably under the circumstances is a **breach of duty** of care and may constitute negligence. **Negligence** is failure to use reasonable care, resulting in damage or injury to another.

Standard of care is defined as the degree of care that a reasonably prudent person should exercise in the same or similar circumstances. In legal matters, the testimony of expert witnesses is often used to establish the standard of care. Also considered is the customary practice in the field. Ultimately, courts determine whether individuals or parties acted in a reasonably prudent manner under a given set of circumstances. If a person's conduct falls below the standard of care, and that breach of duty results in harm, he or she may be held liable for the damages or injuries.

Tree Risk Assessment Manual

As an assessor, you must understand the legal implications of your tree risk assessment. Because you are considered an expert in tree risk assessment, you can be held to a higher standard for recognizing hazards and assessing risk in trees than those who don't hold similar credentials. As long as you apply the current, generally accepted standard of care, then your duty of care has been met. This manual will help guide you through those practices.

Conditions affecting trees change constantly; none of us will ever be able to predict every tree failure. Conducting a tree risk assessment neither ensures nor requires perfection. Risk assessment should, however, ensure that all reasonable efforts have been made to identify the likelihood of failure, the likelihood of impact, and the consequences of failure present at the time of assessment.

Safety

There may be occasions when you, as a tree risk assessor, will be exposed to hazardous conditions. Some examples include assessing trees after a wind or ice storm, when some trees have broken and hanging branches or stems have passed the initial failure point; reviewing a work site in preparation for tree pruning or removal work; or assessing risk of trees that have extensive decay. You must be aware of risk to yourself and others.

When you discover an extreme-risk tree, avoid the target zone as much as possible during the assessment. In adverse weather conditions (such as strong winds, snowfall, or freezing rain), it may be advisable to postpone the assessment, or, if you must proceed, bring a colleague with you and have an exit plan in place for leaving the target zone as quickly as possible, just as in tree felling operations.

Figure 1.8 Maintain a safe distance when assessing trees near power lines during and after storms that may cause branches to contact utility wires. The utility company should be notified if there is risk of electrical contact.

Maintain a safe distance when assessing trees near power lines during strong winds or other conditions that may cause branches to contact utility wires, creating an electrical path through the tree to the ground. If electrical contact is made or is likely, identify the location and contact the utility company. They can dispatch specially trained and equipped personnel to handle the problem safely. Action may be required to keep other people from becoming exposed to the electricity.

Figure 1.9 Perform a pre-climbing inspection before any aerial tree risk assessment.

Module 1 – Introduction to Tree Risk Assessment

CASE STUDY

Assignment: The client requested a risk assessment of the river birch *(Betula nigra)* on the left. The house is on a lake with a 50-foot (15 m) conservation buffer that does not allow the removal of a tree without an arborist report declaring the tree to be high or extreme risk. The birch blocks the view of the lake from the master bedroom. The client has implied that she has a very low tolerance for risk and wants the tree removed.

Targets and Site: The lakeside house is in an affluent neighborhood. Targets include the dock and people who use it, and landscape maintenance workers. The house is not within the target zone.

Conditions: The river birch is chlorotic and has a thin canopy. The tree has codominant stems and a few small-diameter dead branches at the tips of at least three branches. Visual observations and sounding the trunk and buttress roots do not identify any indicators of decay.

Analysis: Using the methodologies in this manual, the risk rating for the river birch was determined to be low. However, during the assessment, you noticed that the white oak *(Quercus alba)* on the right has an open cavity, extensive internal decay, and a vertical crack associated with the cavity opening. The distance from the tree trunk to the house is 0.75 times the tree height. A quick, limited visual assessment of this tree led to the conclusion that it was a high-risk tree.

Recommendation: Crown clean to remove the dead branches from the birch, cable to reduce the risk of failure of the codominant stems, and consider measures (described in a report) to improve health and vigor. Recommend further assessment of the oak.

Provided by E. Thomas Smiley

Tree workers approaching a tree should conduct their own pre-work inspection before entering the target zone. The crew leader needs to consider the position of all ground workers and equipment before anyone enters the tree. Tree risk inspection results should be provided to the crew working on the tree.

Whenever a tree is to be climbed for an advanced aerial assessment, the tree should receive a preliminary risk assessment prior to climbing (pre-climbing inspection). If the tree or tree part cannot support the additional load of the climber and rigging, work should be performed from an aerial lift or crane.

Summary

Tree risk assessment is the systematic process used to identify, analyze, and evaluate tree risk. It is a subset of tree risk management. An understanding of the basic terminology and fundamental concepts of tree risk assessment provides the foundation upon which you will build knowledge and experience. Assessment is one part of the larger issue of tree risk management. As you work through this manual, you will learn about the levels of assessment, and you will become familiar with the methods and techniques used to assess tree risk. You must be aware of the legal, ethical, and safety issues involved. In managing trees, the risk manager should strive to strike a balance between the risk that a tree poses and the benefits that individuals and communities derive from trees.

Key Concepts

1. It is impossible to maintain trees free of risk; some level of risk must be accepted to experience the benefits that trees provide.

2. The two primary approaches to risk assessment are quantitative and qualitative. Each has advantages and limitations, and each may be appropriate with different objectives, requirements, resources, and uncertainties.

3. Prior to beginning a tree risk assessment, it is important to work with the client to define the scope of work.

4. Risk assessment requires a strong sense of professional responsibility. By signing and submitting your assessment report, you are explicitly stating that you have undertaken a thorough and technically valid assessment within the scope of work and defined level of inspection.

5. There may be times when you recognize that you do not have the appropriate skills or equipment to undertake some aspects of the risk assessment process. In such situations you should reserve judgment and, upon approval of your client, bring in a specialist who can undertake those aspects for you.

6. The duty of care for a tree risk assessor is to use the generally accepted standard of care (as defined in applicable standards and best management practices) when evaluating tree risk.

7. There may be occasions when you will be exposed to hazardous conditions when inspecting a tree. You must be aware of risk to yourself and others and take appropriate precautions. When practical, tree risk inspection results should be provided to tree workers prior to climbing.

Levels of Assessment

– Module 2 –

Levels of Assessment

Module 2

Learning Objectives

- Describe a Level 1 limited visual assessment, and explain when it would be used.
- Describe a Level 2 basic assessment, and explain how it differs from a limited visual assessment.
- List the common tools that might be used in a basic assessment.
- Discuss the advantages and limitations of each assessment level.
- Discuss how the limitations of a Level 1 limited visual assessment or Level 2 basic assessment might lead you to suggest a higher level of assessment.
- Describe the various advanced assessment tools and techniques, and explain how they might be used to gather specific information.
- Explain with specific examples why it is important for assessors to understand the possibilities and limitations of the techniques they employ.

Key Terms

advanced assessment	diameter tape	load testing	strength loss
aerial inspection	drive-by	mallet	tomography
aerial patrol	ground-penetrating radar	probe	tree population
basic assessment		resistance-recording drill	walk-by
buttress root	hand pull test		
clinometer	level(s) of assessment	root collar excavation	
decay-detection device	LiDAR	sonic wood assessment	
	limited visual assessment	static pull test	

Module 2 – Levels of Assessment

Introduction

Tree risk assessment can be conducted at different levels of detail, each employing varying methods and providing the client with varied options for reporting and recommendations. The level selected should be appropriate for the assignment, as discussed in Module 1.

As a tree risk assessor, you should explain the options for **levels of assessment** to your clients so that, together, you can establish the appropriate level. Explain that if the tree conditions cannot be adequately assessed at the specified level, you may recommend a higher level or different type of assessment. However, you are not required to provide the higher level if it is not within the scope of the original assignment, without additional compensation, or without modifications to the agreement or contract. The level to be employed should be specified in the scope of work that is established between you and your client(s) prior to conducting an assessment. All levels of assessment use the evidence available at the time of inspection as the basis for the evaluation of targets, site factors, and defects.

The ANSI A300 standard for risk assessment and ISA's *Best Management Practices: Tree Risk Assessment* define three levels of tree risk assessment:

- Level 1: Limited visual assessment
- Level 2: Basic assessment
- Level 3: Advanced assessment

Level 1: Limited Visual Assessment

Level 1 assessment involves a visual assessment of an individual tree or a population of trees near specified targets, conducted from a specified perspective, to identify certain obvious defects or specified conditions. A **limited visual assessment** typically focuses on identifying trees with *imminent* and/or *probable* likelihood of failure (see Module 7).

Typically, in a limited visual assessment, one or two of the three factors used when performing a tree risk assessment (likelihood of failure, likelihood of impact, consequences of failure) is/are considered as a constant. For example, if street trees are being assessed, the targets might be consistent along the entire street, and the likelihood of impact and consequences could be rated the same for each tree. The assessor would focus on the likelihood of failure. In another example, in a Level 1 risk assessment along a utility transmission line, the assessor would predefine the target as the electrical conductor and identify trees that would impact the target if they failed. In this example, the consequence of failure (disruption of electrical services) is held constant.

Limited visual assessments are the fastest, but least thorough, means of assessment and are intended primarily for managing large populations of trees when time and resources are limited. When conducted by trained professionals, limited visual assessments can provide tree managers with an adequate level of information that can accomplish their risk management goals. The assessment is often done on

Level 1: Limited Visual Assessment Process

- Identify the location and/or selection criteria of trees to be assessed.
- Determine the most efficient route for assessing large populations of trees and documenting the route taken.
- Assess the tree(s) of concern from the defined perspective (for example, walk-by or drive-by).
- Record locations of trees that meet the defined criteria (for example, significant defects or other conditions of concern, trees that require mitigation, or trees that require a higher level of assessment).
- Evaluate the risk based on observations and assumptions (a risk rating is optional).
- Submit a report indicating risk level(s) and mitigation options and/or recommendations.

a specified schedule and/or immediately after storms to rapidly assess a **tree population**. Tree inventories often include a Level 1 risk assessment unless a Level 2 risk assessment is specifically called for in the inventory.

In a Level 1 assessment, you would perform a visual assessment from one side or by an aerial flyover, typically looking for obvious defects such as, but not limited to, dead trees, large cavity openings, large dead or broken branches, obvious fungal fruiting structures, large cracks, and severe or uncorrected leans. Only defects visible from the perspective from which the assessment is made can be expected to be identified. Limited visual assessments are sometimes used as a filter to identify trees that should later receive a more thorough assessment.

The client may specify inspection for certain conditions of concern, such as lethal pests or symptoms associated with root decay, which may or may not be considered as part of a Level 1 assessment.

The type of inspection may include one of the following:

Walk-by is a limited visual inspection of one or more sides of the tree, performed as the inspector walks past a tree. You may need to stay on the sidewalk or footpath, on public property, or within a right-of-way. In some cases, you may want to walk around certain trees to gain a more complete perspective, or the scope of work may specify it, although this alone would be insufficient to elevate the assessment to Level 2. The scope of work must be clearly defined to avoid misunderstandings about expected levels of inspection.

Drive-by (windshield) is a limited visual inspection of one side of the tree, performed from a slow-moving vehicle. This type of inspection is often performed by municipalities, councils, utilities, or other agencies or landowners who have large populations of trees to inspect with a limited budget.

Aerial patrol is observation made from an aircraft overflying utility rights-of-way, other areas, or individual trees. This type of inspection is conducted

Figure 2.1 Walk-by is a limited visual inspection of one or more sides of the tree, performed as the inspector walks past a tree.

by some electric utility companies or their contractors to identify threats to the electric transmission system. Unmanned aerial vehicles (UAV) may be used for either Level 1 or Level 3 inspections. Sometimes

Figure 2.2 Drive-by is a limited visual inspection of one side of the tree, performed from a slow-moving vehicle.

Figure 2.3 Aerial patrol inspections are made from an aircraft overflying utility rights-of-way or other large areas.

especially with an aerial (airborne) patrol, it may not be possible to identify all details, including the tree species.

An advantage of limited visual assessments is that they can be used as a relatively quick screening tool for assessing large populations of trees. In some cases, recommended mitigation will be clear (for example, remove the dead tree or provide clearance for the electric transmission line). Other times, you will not be able to make recommendations without further inspection. When trees of concern are identified, you might recommend a basic inspection to provide additional information. The scope of work could address whether recommendations for higher levels of assessment are expected.

A constraint of limited visual inspections is that some conditions may not be visible from a one-sided inspection of a tree. Also, a Level 1 risk assessment may not be adequate to provide risk mitigation options. You might use the Level 1 assessment to determine which trees require further inspection at the basic or advanced levels.

a more detailed, ground-based inspection may be specified to confirm observations. Images may be recorded to document observations.

LiDAR (Light Detection and Ranging), is a remote sensing method that uses laser technology to measure tree size and location in relation to the target of concern. In this case, the assessor may be focusing on the likelihood of impact if a failure were to occur.

When you identify a tree of concern in a limited visual inspection, record information about that tree as specified in the scope of work. At a minimum, this information should include the tree location and any recommended remedial action. The information recorded may also include the species name, tree size, defect or condition identified, and a work priority. Because of the limitations of this level of inspection,

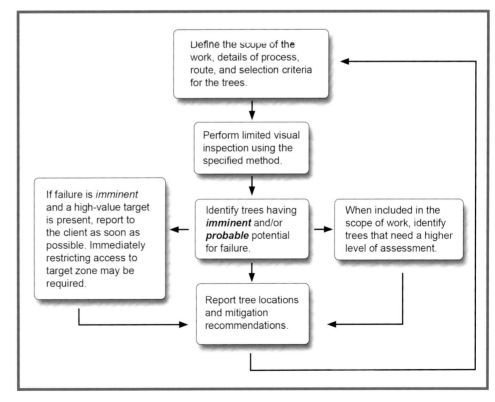

Figure 2.4 Flow chart of Level 1 limited visual assessment procedures.

CASE STUDY

Assignment: Level 1 limited visual assessment, walk-by inspection.

Targets and Site: Urban park, trees along the street. Targets include pedestrians and parked cars.

Conditions: A London planetree *(Platanus × acerifolia)* on the edge of the park by a street has codominant stems with a long, vertical crack between the stems.

Analysis: Failure of either stem could impact multiple targets. Cannot determine the depth of the crack, internal conditions, or the extent of any response growth in a walk-by assessment.

Recommendation: Follow up with a Level 2 basic assessment.

Provided by Julian A. Dunster

Level 2: Basic Assessment

A Level 2 or **basic assessment** is a detailed visual inspection of a tree and its surrounding site, and a synthesis of the information collected. This is the level of assessment that is commonly performed by arborists in response to clients' requests for individual tree risk assessments. It is ground-based and requires you to inspect completely around the tree—looking at the site and at visible buttress roots, trunk, and branches. Look at the tree from some distance away, as well as close up, to consider crown shape and surroundings.

As part of a basic assessment (and without putting it into the category of advanced assessment), you may choose to use simple tools to acquire more information about the tree or any potential defects. However, the use of these tools is not mandatory unless specified in the scope of work.

Measuring tools. You may want to include a **diameter tape**, **clinometer**, or a tape measure to determine tree dimensions.

Binoculars. You can use binoculars to inspect the upper portions of a tree's crown to look for cavities, nesting holes, cracks, weak unions, and other conditions and growth responses. A basic assessment does not include climbing the tree, so binoculars can often provide a closer look without your needing to leave the ground.

Magnifying glass. A magnifying glass (hand lens) may be used to help identify fungal fruiting bodies or pests that may affect the overall health of the tree. Although not every pest or condition found on a tree during a risk assessment will necessarily be relevant, it is important in a basic assessment to assess the growth and health of a tree as part of the analysis of the tree's ability to adapt to future stresses and loads.

Module 2 – Levels of Assessment

Figure 2.5 Sounding. Strike the tree with a mallet and listen for tonal variations that indicate dead bark or internal hollows.

Figure 2.6 Probing. Use a stiff, small-diameter rod to probe into cavity openings to estimate their size.

Mallet. You can use a broad-headed **mallet** made of wood, rubber, nylon, leather, or resin to "sound" the tree. Tap the tree trunk and root flare in multiple places and listen for tone variations that may indicate hollows or dead bark. An internal cavity that has a thin wall will typically produce a deeper, more resonating sound than a solid stem. Dead bark will make a clapping sound when you strike it.

Sounding with a mallet is useful in judging whether and where to perform more advanced assessments. Having a good ear for tonal variations is a necessity, but for most people, practice and experience in sounding techniques will lead to greater proficiency and higher reliability.

Probe. A **probe** is a stiff, small-diameter rod, stick, or wire that is inserted into an open cavity to estimate its size and extent, or to investigate for the presence of decay. You can use a probe anywhere you see an opening in the tree; the resistance met in probing can help you detect decayed wood. However, because not all angles will be accessible and there may be sections of nonfunctional wood adjacent to a cavity, you should consider this assessment technique only an approximation of the extent of decay. Be careful not to injure or disrupt nesting of wildlife living in the cavity, and avoid getting bitten.

Digging tools. A trowel, shovel, or other hand tool can be used to conduct minor excavations to expose surface roots or the root collar. Take care not to damage roots during the excavation process. More extensive root collar excavations are considered an advanced assessment.

Figure 2.7 Excavation. You can use simple tools to conduct shallow excavations to expose the root collar or roots at the soil surface. Take care not to damage roots during the excavation process.

Copyright ©2017 International Society of Arboriculture. All rights reserved.

Tree Risk Assessment Manual

Compass. You may wish to use a compass to determine the orientation of the tree, site, and defects or conditions noted on the tree(s) being assessed. There are many smartphone and tablet applications (apps) that include a compass, as well as functionality to determine height and lean.

Camera. Taking photographs to document conditions that you see is usually a good idea. A camera can also be used to "zoom in" to better observe conditions in the canopy of a tree.

An advantage of a basic assessment over a limited visual assessment is that each tree is inspected more closely, and from all sides, increasing the amount of information and detail you can observe and/or measure. Basic tools may be used to enhance the information collected.

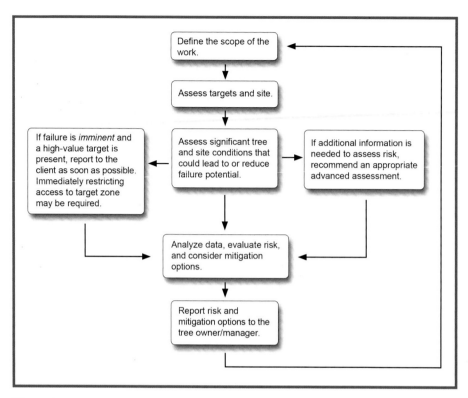

Figure 2.8 Flow chart of Level 2 basic assessment procedures.

Often a basic assessment is adequate for assessing risk and making recommendations, but it sometimes reveals the need for more advanced assessment measures.

Level 2: Basic Assessment Process

- Locate and identify the tree or trees to be assessed.
- Determine the significant targets and target zone for the tree or tree parts of concern.
- Review site conditions and available site history.
- Inspect the tree visually—using binoculars, mallet, probe, or trowel, as desired or as specified in the scope of work.
- Assess expected loads, or changes in loads, on the tree and its parts.
- Assess general tree health.
- Record observations of site conditions, defects and conditions of concern, and response growth.
- If considered necessary, recommend an advanced assessment.
- Analyze available data to determine the likelihood of failure and impact and consequences of failure in order to evaluate the degree of risk.
- Develop mitigation options and estimate residual risk for each option.
- Develop and submit the report/documentation, including, when appropriate, advice on reinspection interval.

The primary limitation of a basic assessment is that it includes only conditions that can be detected from a ground-based visual inspection. Internal, below-ground, and upper-crown factors may be impossible to see or difficult to assess, thus remaining largely undetected or unevaluated.

Assessing Single Trees vs. Assessing Populations of Trees

The processes needed to assess a population of trees are similar to those needed to assess a single tree. Target, site factors, and tree characteristics all need to be evaluated. Typically, sites are grouped when assessing a population of trees, often with a common target or set of targets (e.g., a road, the easement of a power line, or a golf course fairway). An assessment of a population of trees may use a Level 1 assessment as a filtering mechanism to determine which trees or subsets of the population will require a Level 2 assessment. In some cases, where a population of trees is being assessed, the priority may be to first locate any extreme-risk trees and have them singled out for immediate mitigation. Whichever approach you take during a basic assessment, you must be systematic and thorough, making note of all aspects that affect the overall risk rating.

Level 3: Advanced Assessment

Advanced assessments are performed to provide detailed information about specific tree parts, defects, targets, or site conditions. You might choose to conduct an advanced assessment of some type in conjunction with or after a basic assessment if you need additional information and the client approves the additional service.

Specialized equipment, data collection and analysis, and/or expertise are usually required for advanced assessments. These assessments are, therefore, generally more time intensive and more expensive.

Advanced assessments can provide additional information that may make the difference between recommending tree or branch retention or removal. As the risk assessor, you, along with the risk manager, have an important role to play. You will identify additional information needed and recommend appropriate procedures and methodologies, with consideration for what is reasonable and proportionate to the specific conditions and situations.

The risk manager/property owner considers the value of the tree to the owner and community, the possible consequences of failure, and the time and expense required for the advanced assessment, within the framework of laws and regulations.

There are many types of advanced assessments that can be conducted (see box, next page). A few are described in this module. Be aware, however, that all technologies involve some uncertainty. Each technique has limitations that may include accuracy of

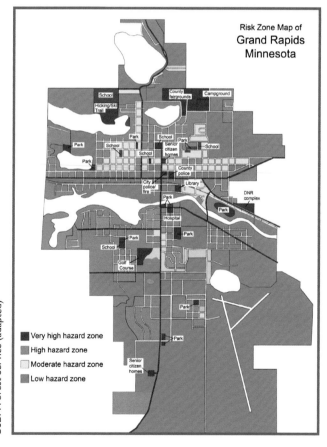

Figure 2.9 Typically, sites are grouped when assessing a population of trees, often with a common target or set of targets (such as a road, the easement of a power line, or a golf course fairway). Assessment of a population of trees may employ a Level 1 assessment as a filtering mechanism to determine which trees or subsets of the population will require a Level 2 assessment.

the measurement, how well the measurement and sampling locations represent the condition, and how the data are interpreted and evaluated with regard to risk. Any evaluation of an individual tree or target will be considered a qualified estimation rather than an accurate measure.

If advanced assessment techniques require equipment and training beyond what you possess, you should seek the assistance of a more qualified person. In addition to the need for technical competency to perform and interpret the assessment, there can be safety issues involved with some advanced procedures. Standard safe work practices and procedures should be applied in all instances.

Aerial Inspection

Aerial inspection (crown inspection) is inspection of the aboveground parts of a tree, especially those not visible from the ground, including the upper trunk and the upper surfaces of stems and branches. Sometimes, defects and other conditions are obscured by climbing plants, mistletoe, moss, or epiphytes, which require an aerial inspection to be revealed.

Figure 2.10 Aerial inspection (crown inspection) involves the aboveground parts of a tree, especially those not visible from the ground.

Level 3: Advanced Techniques

Many techniques can be considered for advanced risk assessment—individually or in combination:*

- Aerial inspection and evaluation of structural defects in high stems and branches
  - visual inspection
  - decay testing of branches
  - UAV photographic inspection
- Detailed target analysis
  - property value
  - use and occupancy statistics
  - potential disruption of activities
- Detailed site evaluation
  - history evaluation
  - soil profile inspection to estimate root depth
  - soil mineral and structural testing
- Decay testing
  - increment boring
  - drilling with small-diameter bit
  - resistance-recording drilling
  - single-path sonic (stress) wave
  - sonic tomography
  - electrical impedance tomography
  - radiation (radar, X-ray)
  - microscopic analysis for fungal type
- Health evaluation
  - tree ring analysis (in temperate-zone trees)
  - shoot length measurement
  - detailed health/vigor analysis
  - starch assessment
- Root inspection and evaluation
  - root and root collar excavation
  - root decay evaluation
  - ground-penetrating radar
- Storm/wind load analysis
  - detailed assessment of tree exposure and protection
  - computer-based estimations according to engineering models
  - wind reaction monitoring over a defined interval
- Measuring and assessing the change in trunk lean
- Load testing
  - hand pull
  - measured static pull
  - measured tree dynamics

* Inclusion of specific techniques in this list should not be considered an endorsement of the use of that technique.

Aerial inspections usually include a visual assessment of defects, conditions, and response growth. Conditions of particular importance include significant branch unions (codominant branches, included bark, or presence of decay), cracks in branches, sunscald on the tops of branches, weak branch unions, and bark damage from bird or animal feeding. In addition, aerial inspections may include evaluation of internal decay.

Aerial inspections can be done from an aerial lift, adjacent building, ladder, unmanned aerial vehicle, or by climbing the tree; you should determine that the tree is safe to climb before entering the tree. Inspection by climbing is likely to be more thorough than from a building, ladder, or aerial device. Visual inspection from the ground using binoculars is not considered an advanced assessment, but it may be part of a basic assessment.

Assessment of Internal Decay

It is often difficult to determine the location and extent of internal wood decay during most basic assessments. When more certainty is necessary, you can get a better estimate using one of several decay-detecting techniques, including drilling and the use of sonic devices. Several **decay-detection devices** are on the market; some have been better researched than others. They differ in precision, accuracy, resolution, and reliability. Moreover, no one piece of equipment provides a complete assessment. Two common, well-established techniques are described here as examples of the general approach.

For all decay-testing methods, it is important to carefully select locations for testing so that the size and configuration of the decay column can be reasonably estimated. Understanding decay patterns will help you determine testing sites.

- Decay in urban trees is often not located in the center of the cross section of a circular stem. Guidelines for assessing the severity of decay are typically based on circular stems with centrally located decay. The use of shell-wall ratios that were developed for single-trunked, forest coniferous trees may not be appropriate. If the decay column is not central, if the tree has a large-diameter trunk, or if the tree is an older, mature, broadleaved tree with large scaffold branches and a wide-spreading trunk, shell-wall formulas may not be applicable.

- Often, urban tree decay starts in damaged lateral roots and enters the stem base from the side. Decay may also progress from pruning wounds into the trunk or branches. The location of decay in the stem cross section is more important in **strength loss** than extent of the decay.

Before conducting any advanced assessment tests, use sounding, probing, and visual assessment to identify the best locations and directions for determining the approximate position and extent of the decayed area. It is useful to know the type of decay expected and its implications before testing. Limitations of decay assessment equipment and models are discussed at the end of this module.

Drilling

Two types of drilling tools are in common use to evaluate the extent of decay: handheld electric drills and resistance-recording drills. Both distinguish between solid and decayed wood by the resistance to penetration as the drill bit moves through the wood.

Figure 2.11 Simple handheld electric drills are primarily limited to detecting advanced stages of decay. Assessing decay using this technique relies in large part on the experience and expertise of the operator.

Figure 2.12 Incipient decay, effectiveness of compartmentalization, and response growth rates may be estimated from profiles created by some high-resolution resistance drills. These profiles can be correlated to certain wood-density levels and interpreted for information about wood condition.

One simple type of drilling device is a handheld electric drill fitted with a long (typically, 8 to 18 inches, 20 to 45 cm), small-diameter (1/8 inch, 3 mm), brad-point tip, full-fluted drill bit fitted with a depth gauge device (earplug, cork, etc.) to identify drilling depth. Detection is primarily limited to voids and advanced stages of decay. Assessing decay using this technique relies in large part on the experience and expertise of the operator, and it does not produce a written record.

Another device is a **resistance-recording drill**, which drives a small-diameter spade bit into the tree. As the bit penetrates the wood, the resistance to penetration is recorded as a graphic profile. This recording can prove to be an advantage in reporting and defending results. With training and experience, an inspector can distinguish solid wood from voids and advanced decay. In some cases, incipient decay, effectiveness of compartmentalization, and response growth rates may be estimated from profiles created by some high-resolution resistance drills. Accuracy tends to be better than using a simple handheld drill because of the physical record, but training and experience are important for interpreting the resulting profiles, especially when no obvious cavity is present.

With either method, take care to avoid unnecessary or excessive tree wounding. Keep in mind that drilling into decay can breach CODIT walls (particularly Wall 4), which may allow decay to spread. The number of drillings should be as few as possible but

Figure 2.13 This tree is being tested using a sonic tomography system.

as many as needed for you to evaluate the extent of decay. In addition, you can stop drilling if you can determine that adequate sound wood has already been confirmed, to avoid drilling through CODIT Wall 4.

Sonic Assessment

Sonic wood assessment instruments send a sound (stress) wave through the wood and measure the time it takes for the wave to travel from the sending point to the receiving point. If a crack, cavity, or decay is present, the sound travels around the defect, increasing the transmission time (time of travel) from the sending point to the receiving point, compared to the transmission time through wood with no defect. The device cannot determine, however, the type of defect (decay, cracks, embedded bark, or cavities) that increased the transmission time.

One type of device measures the transmission time between two reference points, which can be a quick test to reveal the presence of decay between them. A single test may miss even major defects or overestimate a small defect. Cracks that are not perpendicular to the line of testing can be missed or may create misleading results. Tests at additional points are needed to provide a more reliable assessment. Two sets of points, forming perpendicular lines, are considered the minimum by many operators to detect large, centrally located defects.

Sonic **tomography** instruments use measurements between many points to create a two- or three-dimensional (if performed on multiple planes) picture called a tomogram. By comparing the results of all time-of-travel measurements, a computer program constructs a graphic representation of the interior of the trunk. The tomogram mainly illustrates the remaining load-carrying parts of the inspected cross section and cannot necessarily distinguish among cracks, decay, or voids within the damaged areas. The resolution of tomography depends on the number of sensors used on a tree.

Although sonic tomography can yield sophisticated images of internal conditions, there are some limitations:

- The instruments are expensive, and proper instruction is necessary for reliable interpretation of results.
- Results are valid only for the plane(s) tested.
- Just as resistance-drilling profiles must be interpreted in relation to typical specific wood anatomical properties of the inspected species, interpretation of color tomogram pictures also requires training and expertise.

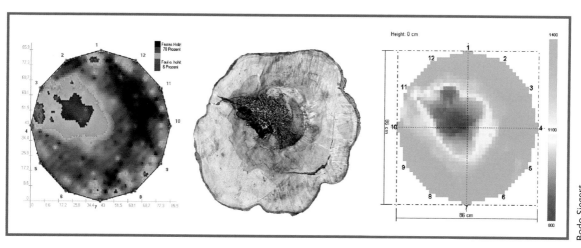

Figure 2.14 Tomograms derived from sonic tomography tests. Different manufacturers portray the results in different ways, but they all help to determine the presence or absence of internal defects. The center image is the actual tree cross section.

In contrast to drilling, sonic devices have substantially less risk of breaching CODIT Wall 4. It should be noted, however, that sometimes sonic tomography interpretations can be enhanced by using resistance drilling to confirm wood characteristics in specific parts of the tree.

Limitations of Decay-Detection Equipment

Many of the instruments used to detect decay are sophisticated and offer accurate and precise results. However, you need to understand where best to apply the tests in the tree and how to properly interpret and apply the results. Evaluations employing results gathered from use of these tools are often far better than those derived from simple visual assessments. However, you should be familiar with the limitations of the tools, techniques, and calculations, as well as the applicability of the results. When appropriate, use international standards to measure, calculate, and represent corresponding results when applying these decay detection methods. Typically, specialized training is needed to properly use the equipment and interpret the data.

Estimating Strength Loss from Decay

There has been relatively little quantitative testing to establish thresholds for failure. Tree failure thresholds that are in use have been established by examining trees that failed and trees that remained standing following storms and comparing the wood loss of each. Research has yet to confirm that any formula accurately represents trunk strength loss, and no critical load threshold has been determined that is valid for all species and aboveground tree forms.

CASE STUDY

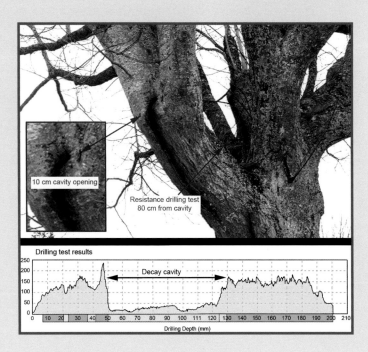

Provided by Julian A. Dunster

Assignment: Advanced risk assessment to determine the extent of decay in a limb.

Targets and Site: Busy residential street and sidewalk.

Conditions: A basic risk assessment of the pin oak (*Quercus palustris*) revealed a lateral limb with a 4-inch (10 cm) diameter hole at the base of a codominant stem. A black streak was noted below the hole. The wood did not sound hollow above or below the hole. The limb was leaning away from the main tree at about a 50-degree angle and out over the street.

Analysis: A resistance drill was used to drill through the limb from the underside to top. Results indicated solid wood on the top and bottom, with the cavity nearly centered and occupying just over one-third of the branch diameter. The drilling also showed a well-defined, strong Wall 4 on the bottom and a weaker Wall 4 on the top.

Recommendation: Retain the limb and monitor. Consider installing a support cable.

Case Update: Twelve years later, the limb was larger, the cavity was covered with woundwood, and there was no external evidence suggesting structural weakness.

Strength Loss Terminology

Formulas to estimate the loss in strength due to decay are in common use. Although we use the terms "strength loss" or "wood loss," these formulas are actually based on the "moment of inertia" of the trunk (its tendency to resist angular acceleration). This level of technical accuracy is not important for most tree risk assessors, though, and references to *strength* are used broadly in this manual.

Figure 2.15 The process of removing materials covering the root collar in order to conduct an assessment is called root collar excavation (RCX). The RCX can be used to locate and examine the main structural roots and to look for decay that could affect stability.

Most of the calculations to assess strength loss due to decay in tree trunks are based on an idealized model of a single, vertical, cylindrical trunk, with the decay centrally located and uniform. When the trunk is leaning, asymmetrical in shape, and the decay is off center, the guidelines for shell-wall thickness recommendations may not apply. Similarly, the extent of the decay column and the adequacy of the remaining shell-wall thickness depend on the trunk diameter. Smaller-diameter stems may be relatively more likely to fail than much larger-diameter stems that have the same percentage of remaining sound shell-wall wood. The height, shape, and mass of the tree crown, as well as the tree's wood properties, affect how much sound wood is required to provide adequate strength. Furthermore, strength has meaning in risk assessment only when assessed relative to the loads that the tree will experience.

Root Assessment

The extent of damage or decay in tree butts, buttresses, and roots is difficult to evaluate in a basic inspection because most roots are beneath the soil surface and their number, size, location, and structure are not visible. Several types of evaluations can be conducted on roots to inspect for decay.

Root Inspection and Evaluation

The simplest method of root inspection is the visual assessment of **buttress roots** at the basal flares or the top surfaces of the roots when exposed. When the roots are not exposed, you would first need to excavate soil or other materials covering the root collar to conduct the assessment. This process is called **root collar excavation** (RCX). The RCX can be used to locate and examine the main structural roots and to look for decay or cuts that could affect stability. This will likely involve excavating the structural roots at least to the point where they become horizontal, and possibly farther. An obvious limitation of this assessment technique is that not all of the major structural roots are located close enough to the surface to be assessed.

The least injurious method of excavation available should be used. This may involve the use of high-pressure air or water. If necessary, hand tools can be used. Take care not to damage the roots or trunk during the excavation process or to destabilize the tree by removing too much soil. After excavation, roots can be inspected for evidence of cutting, injury, decay, response growth, or other conditions.

Root Decay Evaluation

When evaluating root decay, keep in mind that decay in roots may be centralized but often progresses from the bottom of the root upward. Drilling and sonic

Figure 2.16 When evaluating root decay, keep in mind that decay in roots may be centralized but sometimes progresses from the bottom of the root upward.

techniques can help determine the number of roots with decay and the extent of root decay within each root, but they are not designed to quantify the amount of strength loss in the root system. Remember that drilling into decay in roots can also breach CODIT Wall 4, which may allow compartmentalized decay to spread. The number of drillings should be as few as possible but as many as needed to evaluate the presence and extent of decay.

Ground-Penetrating Radar

Ground-penetrating radar (GPR) is a method that uses radar pulses to create images of certain below-ground features. It uses electromagnetic radiation in the microwave band of the radio spectrum to detect reflected signals from subsurface structures or tree roots. The technology is not sophisticated enough to be relied on as a sole decay-detection device, but it can be used in some cases to determine where roots are and whether underground structures might affect root stability. As with many other technologies described, GPR can be expensive and requires training to interpret the results.

Measuring Change of Lean

An increasing angle of lean may indicate an elevated likelihood of failure, especially with changes over a short time frame. Sometimes it is difficult to

Figure 2.17 A digital level or other device can be used to monitor even small changes in lean angle. When trunk angle measurements are made over time, it is important to take them at the same location each time. Note the use of pins to ensure the same placement of the level.

determine if a tree's lean is changing. A digital level or other device can be used to monitor even small changes in lean angle. When trunk angle measurements are made over time, it is important to take them at the same location each time. In addition, digital level readings taken during the dormant season cannot be directly compared with readings taken when the tree has foliage because of the change in load on the trunk. The same is true during times of drought, rain, snow, or ice glazing.

Load Tests

Load tests are used by specially trained tree risk assessors to assist in evaluating the potential for failure, sometimes with specialized measuring devices. There are two available load tests: hand pull test and measured static pull test. Load tests do not attempt to detect internal decay; instead, they use deformation or deflection to detect weakness in the structure and assess the load required to initiate the failure process.

Hand Pull Test

Hand pull testing is most commonly used in a pre-work inspection, but it can be used as part of a tree risk assessment. Because the tree or branch reaction is monitored only visually, you must take care not to overload the tree or tree part to avoid initiating failure. You and all others should be outside the target zone when conducting this type of test.

A **hand pull test** involves installing a line in the tree, and then pulling and releasing the line several times to move the tree or branch. Excessive trunk, codominant stem, root, or soil movement may indicate instability. This test may reveal or enlarge existing crack openings.

Static Pull Test

In a **static pull test**, sensors are attached to the tree to measure outermost fiber strain (stretching and compressing) in the stem or branches, and/or inclination (change in angle) of the root flare in response to a controlled pull. The amount of deformation and inclination, measured by sensors, is compared to reference values for that species to evaluate strength or stability. Working within specific thresholds for tolerable deformations is required to avoid overloading the tree during the load test.

Static tests do not generally provide a good model for the dynamic actions of wind. Research is underway to develop instruments, methods, and interpretation guidelines that would address this concern.

Figure 2.18 It is important not to pull too hard when performing hand pull tests. Using a throwline limits the force that can be applied to trees and branches being assessed by this method.

Figure 2.19 In a static pull test, sensors are attached to the tree to measure outermost fiber strain in the stem or branches and/or inclination of the root flare in response to a controlled pull. The amount of deformation and inclination, measured by sensors, is compared to reference values for that species to evaluate strength or stability.

Other Advanced Assessments

Although the most commonly employed advanced assessment techniques are usually associated with decay detection and strength/stability assessment, there are many other assessments that might be performed. The purpose is to gain more information to refine the assessment of risk. Almost any test, investigation, inspection procedure, or sophisticated calculation that goes beyond the description of a basic assessment could qualify as an advanced assessment. Examples include, but are not limited to, detailed target or site assessments, tree health assessments, laboratory tests to identify decay organisms, and weather data analysis.

Advanced Site and Target Assessments

When more information is needed about a site and its associated targets to either refine the assessment of the likelihood of a target being impacted or to better assess the consequences of failure, a more advanced assessment of the site and/or targets may be conducted. One example is to measure target occupancy rates (such as traffic patterns within or through a target zone) to determine the likelihood of a target being struck in the event of tree or tree part failure. Other examples of target and site analyses are property appraisals (obtained from a qualified appraiser), site history investigations, and soil testing.

Laboratory Analysis

Sometimes it is necessary to collect samples for laboratory analysis. A common example is analysis of fungal fruiting structures to determine the identity of a decay fungus so that an assessment can be made of its virulence and expected rate of spread. Laboratory tests to identify pests, analyze nutrient levels in the soil or foliage, or assess xylem starch levels may be made to evaluate general tree health. This information can be used in determining a tree's ability to compensate for strength loss, compartmentalize decay, and resist stress-inducing factors such as pests.

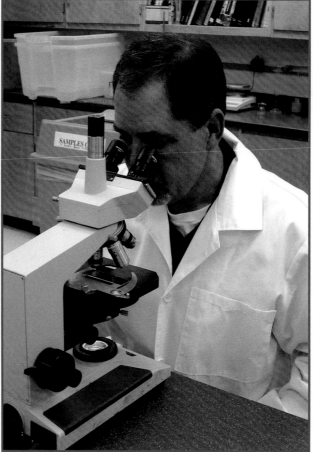

Figure 2.20 Sometimes it is necessary to collect samples for laboratory analysis. An example is identification of decay fungus so that an assessment can be made of its virulence and expected rate of spread.

Weather Analysis

Weather analyses can be used to measure or estimate prevailing wind direction and force or the likelihood of extreme storm events in a given area based on history. A wind rose chart, for example, provides historic information about wind direction frequency and strength. Weather investigations can be particularly important if you are assessing trees in a region unfamiliar to you. Weather analyses can better inform you of the normal frequency of severe loading from wind, snow, ice, or heavy rain.

Summary

Three levels of assessment are defined and available:

- A Level 1 limited visual assessment involves a visual assessment of an individual tree or a population of trees near specified targets, conducted from a specified perspective to identify certain obvious defects or specified conditions.

- A Level 2 basic assessment is the standard assessment performed by arborists in response to a client's request for tree risk assessment. It consists of a detailed visual inspection of a tree and its surrounding site and a synthesis of the information collected.

- Level 3 advanced assessments are performed to provide detailed information about specific tree parts, defects, targets, or site conditions. Advanced assessments can provide additional information that may make the difference between recommending tree retention or removal.

Key Concepts

1. Level 1 limited visual assessments are the fastest, but least thorough, means of assessment and are intended primarily for managing large populations of trees when time and resources are limited.

2. A Level 2 basic assessment requires that you walk completely around the tree—performing a close visual assessment of the site, buttress roots, trunk, and branches.

3. Typically, for populations of trees, the sites are grouped, often with a common target or set of targets. Assessment of populations of trees may employ a Level 1 assessment as a filtering mechanism to determine which trees or subsets of the population will require a Level 2 or Level 3 assessment.

4. Because specialized equipment, data collection and analysis, and/or expertise are usually required for Level 3 advanced assessments, they are generally more time intensive and more expensive than basic assessments. Procedures and methodologies should be selected and applied as appropriate, with consideration for what is reasonable and proportionate to the specific conditions and situations.

5. Almost any test, investigation, inspection procedure, or sophisticated calculation that goes beyond the description of a basic assessment could qualify as an advanced assessment.

Target Assessment

– Module 3 –

Target Assessment

Module 3

Learning Objectives

- Explain what a target is in tree risk assessment, and give examples of types of targets in various sites.
- Discuss how targets are considered when estimating both the likelihood of impact and the consequences of failure in risk categorization.
- Describe how a target zone is determined after assessing a tree's structural condition and likely failure modes, and explain why an individual tree may have multiple target zones.
- Compare and contrast different types of targets and how each affects risk and mitigation options.
- Discuss how to assess target occupancy rates and their effects on risk categorization.
- Describe the relationship between the factors that affect potential impact and the consequences of failure.
- Explain the process used for stratifying and prioritizing targets.

Key Terms

consequences of failure	likelihood of impact	prioritizing targets	target
constant occupancy	mobile target	protection factors	target value
degree of harm	movable target	rare occupancy	target zone
disruption	occasional occupancy	static target	
frequent occupancy	occupancy rate	stratifying targets	

Introduction

In tree risk assessment, the likelihood of a tree failure impacting a target is the combination of the likelihood of tree failure and the likelihood of the failed tree or tree part impacting a target. **Targets** are people who could be injured, property that may be damaged, or activities that could be disrupted by a tree failure. Examples of targets include people, buildings, infrastructure, power lines, vehicles, and landscape structures. The type of target directly affects both the likelihood of impact and the consequences.

Risk assessment is undertaken where there are one or more targets of concern, and the owner/manager thinks that **target value** justifies the expense of assessment. Sometimes, risk assessment is required by law. Evaluating potential targets is one of the first steps in the risk assessment process, whether considering one individual tree or a population of trees. Once you have identified potential targets, you will evaluate each one to assess the likelihood that tree failure will damage it and decide what the likely consequences would be. A pragmatic assessment is needed to assess how likely it is that injury to people or damage to property will occur, rather than assuming that the worst-case scenario will always occur.

The Target Zone

The **target zone** is the area in which the tree or tree part is likely to fall when it fails. When determining the target zone, you should consider the direction of fall, the height of the tree, crown spread, slope of land, wind, potential for dead branch shattering, and other factors that might affect spread of debris. For a whole tree that is growing on level ground and not leaning, the target zone is generally defined by a circle around the tree with a radius equal to at least the tree height. Because falling trees can break up and scatter debris, a radius of up to 1.5 times the tree height is used in some cases. Trees with asymmetric crowns or leans may have irregularly shaped target zones. Trees growing on a slope will typically have a target zone radius greater than tree height on the downhill side of the tree. The target zone for a branch is the area in which the branch could strike.

The target zone for dead trees or trees with dead or brittle branches is generally larger than those with live, flexible branches because dead and brittle branches are more likely to shatter when they hit the ground and to spread debris some distance beyond the tree. In some situations, the target zone also may be larger when the failure of one tree could cause the failure of others. Conversely, the target zone may be reduced in size by adjacent larger trees, strong lower branches, or other factors that will not allow the tree or branch to fall in a given direction.

Trees can fall in unusual ways during adverse weather, striking outside what normally would be considered the target zone. The direction of tree failure is often more related to wind direction than the location of the defect on the tree. But wind direction in a storm is often unpredictable and should not be equated with the direction of the prevailing wind.

What Is a Target?

The most important target is a person. The greatest risk occurs in situations in which many people are unprotected within the target zone for long periods

Figure 3.1 The target zone is the area where a tree or branch is likely to land if it were to fail. Due to the brittle dead branches at the top of this tree, the target zone radius is equal to 1.5 times the tree height.

of time, especially during storms. Areas where people congregate include streets, parking areas, patios, and athletic fields. Although damage to structures is possible, the bigger concern is the people who use them, even though people may not be present at the time of the assessment.

For example, normally you would not identify a sidewalk as a target, but you should consider the people who use the sidewalk. Similarly, when assessing the risk of a tree failing in a parking area, you should consider the parking area's volume of use, possible injury to people who drive the cars, and possible damage to cars that may be parked.

Structures make up another large type of target. Occupied structures are the most important because of the inhabitants, yet all structures that have value and importance to their owner and are within striking range of the tree should be considered. Examples of structures include buildings, garages, storage sheds, decks, and fences.

Multiple targets are present at most sites. For instance, the possible consequences of a tree falling across a road include:

- Damage to a moving or stationary vehicle
- Injury to driver and/or passengers
- A moving vehicle running into the tree that has just failed
- Crashes associated with a vehicle suddenly stopping or swerving to avoid the fallen tree
- Blocked access to points farther along the road
- Disruption of utility services
- Costs and inconvenience factors associated with clean-up and remediation

Targets of particular concern to electrical utility risk managers are electric lines and facilities. Lines brought down by a falling tree are a threat to people who may make contact with them, and there may be a risk of fire. There are also costs, inconvenience, and possible losses associated with the disruption of service that will potentially impact many people. Trees that fall on utilities can pose a serious risk of disruption, in addition to the risks of injuring people or damaging property.

Types of Targets

Known targets are those that are visible to you and those that you have been told about. There may be unknown targets as well, and you need to examine the site with that in mind. For instance, there may be targets that are not present at the time of the assessment and are not readily foreseeable. For this reason, it is best to evaluate the targets in consultation with the client so that site uses can be discussed and unknown targets may be included. The scope of work should define targets of concern and often may consider only high-value targets (for example, people, houses, and power lines).

Targets can be categorized by their ability to move or be moved.

- A **static target** is one that is fixed or not readily moved. Examples include buildings, utility facilities, or other structures.

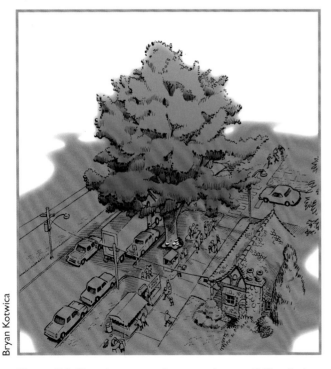

Figure 3.2 Targets are people, property, or activities that could be injured, damaged, or disrupted by a tree failure. If this large tree were to fail at the root system, targets would include the pedestrians, food vendors, traffic, parked cars, power lines, and the house.

Module 3 – Target Assessment

Figure 3.3 Static or unmovable targets include permanent or semi-permanent (not easily moved) structures such as buildings and power lines.

Figure 3.4 Movable targets can move or be moved relatively easily. They may be relocated beyond the target zone, however, movable targets might later be moved back into the target zone if not affixed to the ground.

- **Movable targets** are those that can be relocated, such as sculptures, information boards, picnic tables, parked cars, or playground equipment.
- A **mobile target** is one that is in motion or intermittently moving. Examples include people (such as pedestrians or bicyclists) and vehicles on a sidewalk, footpath, or road.

The type of target affects not only the risk assessment but also the way in which identified risks can be mitigated.

Static Targets

Static or unmovable targets include permanent or semi-permanent (not easily moved) structures such as buildings, motorways (interstate highways), and power lines. They also include structures such as park benches that are permanently fixed to the ground and cannot be easily picked up and moved.

Movable Targets

Movable targets can move or be moved relatively easily. Parked cars, park benches, picnic tables, and portable toilets are all movable targets. Movable targets may be relatively simple to relocate to a place beyond the target zone, thus reducing the risk level. If there is a chance that the movable target will later be moved back into the target zone, however, it may be reasonable to rate it at a higher level until the manager can ensure it has been permanently relocated. For example, people often move picnic tables under trees so that they will be in the shade and thus could expose themselves to greater risk without realizing it.

Mobile Targets

Mobile targets are those that are moving continuously or intermittently. People move through target zones either as pedestrians or as occupants in a vehicle. Because mobile targets move at different speeds and sometimes stop for varying periods of time, assessment of mobile targets must include

Figure 3.5 Mobile targets are those that are moving continuously or intermittently.

consideration of the amount of time they are within the target zone (the occupancy rate).

Occupancy Rates

The amount of time one or more targets are within the target zone—its **occupancy rate**—is the primary component of assessing the likelihood of a target being impacted. Not all targets may be present in the target zone at all times. Occupancy rates can be classified as constant, frequent, occasional, or rare.

Static targets, by definition, represent a constant occupancy, while movable and mobile targets can be in any of the following four classifications.

Constant Occupancy

Constant occupancy indicates that a target is present at nearly all times, 24 hours a day, 7 days a week. This can be the case for static, movable, or mobile targets. A static, unmovable target such as a building is an example of constant occupancy. If there is a steady stream of mobile targets moving through the target zone, this can also be classified as constant occupancy. An example would be a high volume of traffic along a street or highway. Each vehicle or person walking may occupy the target area for a very short time but, in aggregate, they represent constant occupancy.

Frequent Occupancy

Frequent occupancy indicates that the target zone is occupied for a large portion of the day or week. Suburban streets with moderate traffic volume, car parks for facilities that are open during the daytime only, sidewalks in shopping areas, and busy delivery areas are examples of frequent occupancy.

Occasional Occupancy

Occasional occupancy describes a site that is occupied by people or other targets infrequently or irregularly. Examples include country roads, low-use footpaths, and low-use sections of parks. In some instances,

Figure 3.6 Constant occupancy indicates that a target is present at nearly all times, 24 hours a day, 7 days a week. Static, unmovable targets such as buildings or power lines are examples of constant occupancy.

Figure 3.7 Frequent occupancy indicates that the target zone is occupied for a large portion of a day or week. Suburban streets that receive moderate volumes of traffic, car parks for facilities that are open only during the daytime, sidewalks in shopping areas, and busy delivery areas are examples of frequent occupancy.

Module 3 – Target Assessment

Figure 3.8 Occasional occupancy describes sites that are occupied by people or targets infrequently or irregularly. Examples include country roads, low-use footpaths, and low-use sections of parks.

Figure 3.9 Rare occupancy applies to sites that are not commonly used by people. Backcountry trails, fenced areas that are well away from more actively used parts of a site, remote parts of an estate, and gardens through which neither workers nor visitors typically pass would all have rare occupancy.

a seldom-used area may be heavily used for short periods. Examples might include cemeteries, a field surrounded by trees that is used for special event parking, or trails and access roads used only when an event is underway. The client or tree manager may define whether the risk assessment will consider low- or high-use times, or both.

Rare Occupancy

Rare occupancy describes sites that are not commonly used by people or other mobile/movable targets. Backcountry trails, fenced areas that are well away from more actively used parts of a site, remote parts of an estate, and gardens through which neither workers nor visitors typically pass would all have rare occupancy. The client or tree manager may decide, as a matter of policy, that the risk in these areas is so low that risk assessment is not justified.

Occupancy Rates of Mobile Targets

Moving targets can be complicated to assess because their presence can vary. For example, pedestrian traffic along a city street may be infrequent or rare during night hours but frequent during business hours.

The exposure time of mobile targets in the target zone depends on factors such as how fast the target is moving, whether a target is likely to be stationary for a period of time within the target zone, and the size of the target zone. For example, if vehicles or people move past the tree along a predetermined route such as a trail or road, they spend only a short time in the target zone. At a typical walking pace of 3 miles per hour (5 km per hour), the exposure time is on the order of 5 to 10 seconds, depending on the size of the target zone. If, however, a pedestrian stops to wait for a traffic light or read a sign, the exposure time will be longer. As you assess the target zone, you should consider factors that could affect target exposure time.

A passing car has a lower exposure time than a pedestrian because it moves at greater speed. However, drivers of vehicles need to see a falling tree well in advance in order to avoid hitting it. If many people or vehicles are passing by the tree, then the cumulative exposure time will be higher, and the risk of harm will increase.

In some cases, you may want more information about target occupancy to assess tree risk. Although beyond the scope of a basic assessment, it may be possible to obtain traffic data and use patterns by time of day or season and to assign some level of priority based on known use patterns. Even in places where there are credible traffic data, the volume of traffic varies by time of day (for example, during rush hour versus the middle of the night).

Weather conditions may further influence occupancy rates. Most tree failures occur during adverse weather events. In general, it is reasonable to assume that there will be fewer people occupying a park, trail, or pedestrian area during torrential rains, typhoons, hurricanes, tornadoes, blizzards, or ice storms. Conversely, in many cities, car traffic increases in rainy weather as people avoid walking, bicycling, or using public transportation if it involves exposure to the weather. It is also important to consider outdoor areas where people will gather during storm events.

Targets and Likelihood of Impact

One of the factors that must be considered in tree risk assessment is the likelihood of a failed tree or tree part impacting a target of concern. To estimate this likelihood, you estimate, research, or measure the occupancy rate of any targets that would be impacted by the failure (the target zone) and any factors that could protect the target from impact of the falling tree or tree part. For example, a target may be directly under a broken branch but protected from impact by other trees or man-made structures. In this example, the occupancy rate might be constant, but the **likelihood of impact** might be very low because of the **protection factors**. The likelihood of impacting a target can be categorized using the following guidelines:

- **High:** The failed tree or tree part is likely to impact the target. This is the case when there is a constant target, with no protection factors, and the direction of fall is toward the target.

- **Medium:** The failed tree or tree part could impact the target, but is not expected to do so. This is the case for people in a frequently used area when the direction of fall may or may not be toward the target. An example of a *medium* likelihood of impacting people could be passengers in a car traveling on an arterial street (frequent occupancy) next to the assessed tree with a large, dead branch over the street.

- **Low:** There is a slight chance that the failed tree or tree part will impact the target. This is the case for people in an occasionally used area with no protection factors and no predictable direction of fall; a frequently used area that is partially protected; or a constant target that is well protected from the assessed tree. Examples are vehicles on an occasionally used service road next to the assessed tree or a frequently used street that has a large tree providing protection between vehicles on the street and the assessed tree.

- **Very low:** The chance of the failed tree or tree part impacting the specified target is remote. Likelihood of impact could be *very low* if the target is outside the anticipated target zone or if occupancy rates are rare. Another example of *very low* likelihood of impact is people in an occasionally used area with protection against being struck by the tree failure due to the presence of other trees or structures between the tree being assessed and the targets.

Usually targets are assessed on an individual basis, but they can also be combined to provide the client with a better perspective of what would happen in case of a tree failure. As discussed previously, this is typically done for moving targets such as vehicles on a road or pedestrians on a sidewalk. You might also consider this approach for multiple structures under or near a large, wide-spreading tree. Rather than consider each structure individually, you might define the target as "structures" and consider the likelihood of any of them being impacted. Combining targets is likely to increase the likelihood of impact, possibly enough to raise the rating.

Targets and Consequences of Failure

The consequences of a tree failing and impacting a target are a function of the value of the target and the potential injury, damage, or **disruption** (harm) that could be caused by the impact of the failure. The amount of damage depends on the part size, fall characteristics, fall distance, and any factors that may protect the target from harm.

Four categories of **consequences of failure** are used in this risk assessment method: *severe*, *significant*, *minor*, and *negligible*.

Severe consequences are those that could involve serious personal injury or death, high-value property damage, or major disruption of important activities. Examples of severe consequences include:

- injury to one or more people that may result in hospitalization or death
- destruction of vehicles of extremely high value
- major damage to or destruction of a house
- serious disruption of high-voltage distribution circuits or transmission power lines

Significant consequences are those that involve substantial personal injury, moderate- to high-value property damage, or considerable disruption of activities. Examples of significant consequences include:

- injury to a person requiring medical care
- serious damage to a vehicle
- high monetary damage to a structure
- disruption of distribution primary voltage power lines
- disruption of arterial traffic that causes an extended blockage and/or rerouting of traffic

Minor consequences are those that involve minor personal injury, low- to moderate-value property damage, or small disruption of activities. Examples of minor consequences include:

- minor injury to a person, but not requiring professional medical care
- damage to a landscape deck
- moderate monetary damage to a structure or vehicle
- short-term disruption of power on secondary lines, street lights, and individual services
- temporary disruption of traffic on a secondary road

Negligible consequences are those that do not involve personal injury, involve low-value property damage, or disruptions that can be replaced or repaired. Examples of negligible consequences include:

- striking a person, causing no more than a bruise or scratch
- damage to a lawn or landscape bed
- minor damage to a structure, requiring inexpensive repair
- disruption of power to landscape lighting
- disruption of traffic on a neighborhood street

When evaluating consequences, consider the size of the tree or tree part that could fail and how it could impact a target. If a 4-inch (10 cm) diameter branch falls on a house from a height of 10 feet (3 m) above the roof, the degree of damage would be low, and no injury to people inside would be expected. If the same size branch were to fall from near the top of a large tree with no branches in between to slow it down, more extensive damage could occur. If the lower branches of the tree slowed or stopped the fall of the falling branch, the anticipated damage would be less. Assessing consequences from target value and **degree of harm** is discussed in Module 7.

In estimating how much damage could occur from a tree failure, consider the relative amount of force with which it is likely to strike the target. A falling tree or branch will gain speed as it falls. So, in general, the higher the distance from which a branch falls or the greater the distance from the tree to the target, the greater the force that the tree or branch will have at the point of impact. If the distance from a tree trunk to a well-built, multi-story house is short, a tree that fails may simply lean against the house, causing minor damage. On the other hand, if the

Figure 3.10 If a 4-inch (10 cm) diameter branch falls on a house from a height of 10 feet (3 m) above the roof, the degree of damage would be low, and no injury to people inside would be expected. If the same size branch were to fall from near the top of a large tree with no branches in between to slow it down, more extensive damage could occur.

Figure 3.11 A falling tree will gain speed as it falls. In general, the greater the distance from the tree to the target, the greater the force that the tree will have at the point of impact.

distance is such that the tree can accelerate significantly before the trunk strikes the house, damage may be much greater. If the tree has lower branches that are likely to slow or stop the fall of the trunk, damage may be lessened. In this example, the lower branches serve to protect the target. Large-diameter, wide-growing branches that are low on the trunk also may affect the fall pattern of a tree. If the branches contact the ground well before the trunk does, the fall may be slowed or stopped, or the tree may roll.

Adjacent trees can also influence the consequences of failure. Trees that stand between the tree being assessed and a target may serve to either decrease or increase the target zone of the tree being assessed. If a large tree falls against another tree, the force of the impact may cause one or more sequential tree failures (a domino effect). This is more likely with wet or saturated soil and species prone to root or soil failures. Sequential, whole-tree failure may also occur if newly exposed trees are not well adapted to the lack of shelter they have since the adjacent tree failed.

Protection against falling trees or branches may be provided by structures that surround the people in the target zone. Substantial building structures provide protection against injury to people inside them. Pedestrians, children on playgrounds, and people in tents have no significant protection from falling tree parts; therefore, they can be seriously injured by small- or medium-sized branches.

Module 3 – Target Assessment

CASE STUDY

Assignment: Assess the risk of the black locust (*Robinia pseudoacacia*) for a one-year time frame. The client asked for a second assessment after plans changed the targets.

Targets and Site: Originally, the building in the background was empty and scheduled for demolition. The right stem would fall into a patch of brambles—no target. The left stem would fall onto the grass, but the only potential target was the person who cut the grass, and the occupancy rate was rare, so there was a remote chance of the target being struck. The building was recommissioned for a daycare, however, with plans for a play area on the left side of the tree up to the base of the tree.

Conditions: The tree has two stems that have split apart nearly to the base, but the tree has been that way for many years, and there is some response growth on both sides and at the base. Both trunks have visible decay.

Analysis: Initially, the building was not considered of high value, and there were no other targets of concern. The new plans for the building significantly changed the targets and the consequences of failure. The tree did not change, nor did the likelihood of failure (*probable*). But the likelihood of impact increased from *very low* to *medium* for people using the site, the targets of greatest concern. Consequences of failure changed from *negligible* to *severe*. With a likelihood of failure and impact category of *somewhat likely*, and *severe* consequences of the failure, the risk was rated as moderate.

Recommendation: Several options were presented to the client:

1. Crown clean to remove dead and dying branches, reduce the crown by 20% on the left side that would be over the proposed play area, and cable and brace the two trunks together to reduce likelihood of failure.
2. Build the play area in another location.
3. Remove the tree.
4. Do nothing to the tree, and accept the risk.

The client had a low threshold for risk, given the nature of the targets, and opted to remove the tree.

Provided by Julian A. Dunster

Figure 3.12 Protection against falling trees or branches may be provided by structures or other trees. Substantial building structures provide protection against injury to people inside them.

Vehicles provide some protection against small-branch failure. However, vehicles struck by a large branch or trunk account for a substantial number of tree-related fatalities. Because of this, when evaluating targets, consider traffic flow, speed, street configuration, and occupancy rates within the target zone near roads and parking lots.

Stratifying and Prioritizing Targets

Stratifying (systematic grouping) is a process for classifying targets, and **prioritizing** is a process for ranking targets, according to importance or value. Often, site areas and targets are ranked as high, moderate, and low use or value. Using this process can be helpful when assessing large populations of trees that have a wide range of targets or when assessing an individual tree that has many targets.

One method of stratifying target areas when conducting risk assessment is to consider management levels. For a municipality, stratification may have already been done by the risk management department as part of a larger emergency preparedness policy. There are no absolute rules but, typically, the highest-priority target areas are the following:

- Critical locations such as hospitals, fire and police stations, and ambulance centers
- Main access routes and transportation sites such as airports, harbors, and train stations
- Main arterial roads linking these places

In the event of a severe storm, these areas become critical sites and routes that must be kept open.

Stratification at a local or neighborhood level would typically establish main roads, busy intersections, bridges, bus shelters, pedestrian crossings, and areas that have constant or frequent use as high-priority areas. Shorter inspection intervals may be justified in these areas to identify trees with a higher likelihood of failure and impact.

Site-specific land uses such as parks or golf courses can also be stratified by priority. Questions that you should ask include "Where are cars parked?"

Figure 3.13 Stratification at a local or neighborhood level would typically identify high-priority areas as main roads, busy intersections, bridges, bus shelters, pedestrian crossings, and areas that have constant or frequent use.

Figure 3.14 Site-specific land uses, such as golf courses, can be stratified and prioritized. The heaviest use areas are the access roads, parking lots, clubhouse, and the main paths to and from the course (high frequency, low exposure time). Tees and greens are areas where people congregate less often but for longer periods of time (lower frequency, higher exposure time). Cart paths and fairways have even less use (lower frequency, lower exposure time).

"Is there a concession stand or toilet facility where people congregate for longer periods of time (higher occupancy rates)?" "Where will people congregate during a storm?" These questions can lead to identification of areas with the greatest occupancy rates.

Using golf courses as an example, the heaviest use areas tend to be the access roads, the parking lot, the clubhouse, and the main paths to and from the course (high frequency, low exposure time). The next frequency level down would be areas where people stay for periods of time (lower frequency, but higher exposure time) such as tees and greens. Cart paths and fairways would be a lower rating (lower frequency, lower exposure time) because the number of people is usually lower than the overall number coming and going to places like the clubhouse. All of these aspects determine how targets are assessed.

There may be instances when you are asked to review an entire park or other land-use area and to check all the trees, regardless of targets. Suppose the assignment is to undertake a basic assessment of all trees along the trails, access roads, and all facilities within a regional park. In such a case, broad land-use patterns can be used to estimate occupancy rates.

Once the assessment is complete, priorities for action can then be based on usage rates. For example, if several trees are likely to fail close to a concession stand, and others are scattered throughout the more remote trails, you should note this in the report so that the risk manager can take immediate action to deal with the trees most likely to cause harm.

Targets can be ranked even when assessing only one tree. In some cases, there may be only one or two targets of concern. In those cases, you should evaluate each target and the consequences of failure to it. In other cases, there may be many targets, and it may not be feasible to evaluate every one of them. In those cases, prioritize and evaluate the highest-value targets. For example, if the target zone encompasses a sidewalk, parking area, a fence, and a trellis, it may be adequate to evaluate only the people (using the sidewalk) and vehicles (occupying the parking area) as targets.

When evaluating a population of trees, there may be differences in use patterns that will affect your recommendation for inspection level and frequency. In these cases, it may be beneficial to stratify the targets and evaluate areas of greatest concern more frequently or at a higher level.

It is often said that without a known target, a tree poses no risk, but that does not mean that a failure would have no consequences. To begin with, any tree worker represents a potential target that may be unknown to the assessor. Furthermore, trees and the flora and fauna that surround them have value. There may be environmental benefits associated with the tree, such as a wildlife habitat, which the manager feels warrant additional work to maintain the tree in place. In such instances, the possible damage and loss of value is the tree itself rather than anything it might strike in the event of failure. Therefore, there are always some consequences associated with failure, though risks to people or structures may not be a consideration.

Summary

In tree risk assessment, targets are people who could be injured, property damaged, or activities disrupted by a tree failure. Risk assessment is undertaken when there is a target of concern. Understanding the importance of the target is a key part of risk assessment, whether for one tree or many trees. The targets must be identified and the potential consequences evaluated. The type of target has a direct connection to the likelihood of impact and depends on the occupancy rates. Once these aspects have been considered, the next step will be to examine the site to see how they influence the potential for tree failures.

Key Concepts

1. The target zone is the area in which the tree or branch is likely to strike when it fails. For a whole tree that is growing on level ground and not leaning, the target zone is generally defined by a circle around the tree with a radius of 1 to 1.5 times the tree height.

2. The most important target is a person. The greatest risk occurs in situations in which many people are unprotected within the target zone for long periods of time, especially during storms.

3. Determining the type and importance of targets will help establish the amount of time and money needed for risk assessment and mitigation.

4. A static target is one that is fixed or not readily moved. Movable targets are those that can be relocated, such as sculptures, information boards, picnic tables, parked cars, or playground equipment. A mobile target is one that is in motion or intermittently moving.

5. Consequences of failure are a function of the value of the target and the amount of injury, damage, or disruption (harm) that could be caused by the impact of the failure. The amount of damage depends on the part size, fall characteristics, fall distance, and any factors that may protect the target from harm.

6. Stratifying (systematic grouping) is a process for classifying targets, and prioritizing is a process for ranking targets, according to importance or value. Often, site areas and targets are ranked as high, moderate, and low use or value. Using this process can be helpful when assessing large populations of trees that have a wide range of targets or when assessing an individual tree that has many targets.

Site Assessment

– Module 4 –

Site Assessment

Module 4

Learning Objectives

- Recognize and discuss how a site influences the growth and development of the trees present.
- Identify and describe the key site factors that influence the likelihood of tree failure.
- Recognize weather conditions that affect trees on a site.
- Assess and discuss the interactions among past site changes, tree growth, and weather patterns, and anticipate how these might affect the short- and long-term stability of the trees.

Key Terms

codominant	hydrology	soil compaction	wind load
dominant	interior tree	soil depth	wind velocity
drag	land disturbances	subdominant	windthrow
edge tree	land-use history	suppressed	
forest stand	open-grown	topography	
grade changes	precipitation	wind exposure	

Module 4 – Site Assessment

Introduction

Site factors can have a significant influence on both the likelihood and consequences of tree failure. Changes or disturbances of the site, recent or not, may increase or decrease the likelihood of a tree failing and may affect the likelihood of that failure impacting a target.

In the preceding module, we learned to recognize and evaluate targets. Now we turn to the site itself to identify the site factors that may affect tree stability and the changes or disturbances that may have already affected the trees being assessed.

As a tree grows, it adapts to its environment. Growth and development of a tree are, in part, a response to the forces and conditions the tree has experienced throughout its life. When conducting a risk assessment, we consider weather patterns and site history and estimate how past site changes might affect the tree and its likelihood to fail.

For example, past tree failures (whole tree or tree parts) can provide information about potential site problems. Multiple tree failures in an area might indicate shallow, saturated soil; compaction; or root disease in the area. Alternatively, a wind storm may have taken down many trees and left others standing but unstable.

New land development is another critical factor because it may increase site use (occupancy rate). Stability-related effects of new development include root damage, soil disturbance, altered soil hydrology, and new wind patterns.

All these changes may affect tree health and likelihood of failure. You can learn to read the landscape to identify aspects that are relevant to the assignment, and then interpret them as part of your assessment. Understanding the site will help to inform your observations as you move from the broader context of the site down to tree-specific details. Both negative and positive influences should be noted since both will play a role in assessing risk levels.

Aspects of the Site to Consider

Site examinations can include identification and assessment of the following:

- Patterns of previous tree failures
- Terrain (slope and aspect of the landscape)
- Wind patterns (especially changes) across the larger landscape and at the tree
- Soil characteristics
- General and site-specific drainage patterns
- **Land disturbances** such as flooding, **grade changes**, or altered hydrology
- Construction damage
- Restricted root growing conditions
- General forest characteristics (species, age class, conditions)
- **Land-use history** and changes that have occurred during the tree's lifespan
- Factors that could affect tree health

The goal of site assessment is to identify the site factors that may affect tree stability. In addition, changes or disturbances that may have already affected the trees being assessed can be identified. It can be helpful to start with a broad perspective of the site. You can get a sense of the environmental conditions on the site by simply walking or driving by slowly to observe the larger landscape and its context. The use of aerial photographs can be a helpful first step toward gaining some sense of the site context and adjacent land-use patterns. If a series of images is available over many years, it may be possible to see changing land uses and their effects on trees. Historical images of some areas are available online.

Try to estimate how long ago disturbances occurred—the past year, several years ago, or decades ago. The longer the time lapse, the more time the tree will have had to adapt or respond to the change.

Table 4.1 Examples of possible site analysis activities for the three levels of assessment: limited visual, basic, and advanced.

Level 1: Limited Visual	Level 2: Basic	Level 3: Advanced
Observe overall topography.	Examine detailed topography—macro (area-wide) and micro (in vicinity of tree).	Measure and map site elevation levels and slopes.
Observe new forest edge.	Estimate change in tree exposure to wind.	Measure wind speed variation and map patterns.
Observe gross site changes (e.g., new structures or new grades).	Examine visible site changes (e.g., look for recent excavation or changes in land surface treatment).	Inspect (excavate) areas of site change to determine extent of root damage.
Observe adjacent tree conditions.	Look for similarities or major differences in conditions by species, age, and site-specific locations.	Identify, test, and confirm cause of differences in condition.
Observe prior tree failures (whole tree or component parts).	Identify patterns of whole tree failure and cause (windthrow, root disease, soil depth, soil moisture) or component failures.	Determine specific cause of failures, such as species of root disease.
Observe indications of significant changes in soil moisture/drainage within natural sites.	Examine tree base for J trunk shape (creeping soil) or soil cracking.	Use soil probe or excavate to identify depth, layers, moisture history, and root depth.
Observe obvious vegetation types and health of individual trees and stands.	Examine woody and herbaceous species and health for site information.	Sample and analyze soil to identify compaction, pH, salinity, and toxic substances.

History of Failures

Past failures can sometimes provide information to help assess likelihood of failure. Nearby trees may have experienced similar weather patterns and site changes, and observations of previous failures may be indications of site issues to be examined. Look for signs of past failures such as fallen trees and stumps, broken or hanging branches, broken stubs, scars from branch failures, leaning trees, signs of root breakage, and roots partially pulled from the soil. Look also for signs of pruning, support system installations, or other mitigation measures that might have been taken to reduce risk in the past. When tree failures are noted, look for possible causes such as root disease, waterlogged ground, shallow soil, and construction injury above and belowground.

Weather

Most tree failures occur during periods of adverse weather—wind or ice storms, blizzards, or heavy rains coupled with strong winds. Tree risk assessment is undertaken considering normal circumstances and typical weather conditions, which may include storms. As an assessor, you should be familiar with the typical weather patterns for the region.

Wind

Many forces can cause tree failure, and wind is the most common of these. Knowledge of regional and local wind patterns and their potential interactions with specific site conditions can be important in assessing the likelihood of failure. In some locations where strong winds are rare or infrequent, the wind speed at which failures of structurally sound trees occur may be lower. For example, in some tropical

Module 4 – Site Assessment

Figure 4.1 When wind speeds exceed strong gale force, even defect-free trees can fail.

Figure 4.2 The manner in which wind moves around a tree will depend on several factors, including the height of the tree, and the shape, extent, and openness or porosity of the crown.

areas where gale-force winds are rare, trees may fail at wind speeds that would not damage most wind-adapted temperate-zone trees. Also, the prevailing wind direction may differ from that of the storms that typically cause the most damage. Trees adapt to their locations and to the wind speeds and direction that commonly occur.

Except for sudden branch drop, tree failures in normal wind speeds are usually associated with serious, uncorrected, or unmitigated structural defects or other conditions, alone or in combination. When wind speeds exceed strong gale force (about 50 mph or 80 km/h), even defect-free trees can fail.

Site Factors to Evaluate

- Evidence of past tree failures (branch, trunk, root, and soil) and possible cause (root disease, waterlogged ground, etc.)
- Wind exposure, including *changes* to wind patterns or exposure
- Recent exposure of forest trees due to clearing for new development or storm damage that has opened up a new edge among a group of trees
- Site changes that may have altered wind and/or sunlight exposure
- Evidence of root damage such as trenching, excavation, filling (placement of fill materials or alteration of native soil grades), compaction, and any other construction activities
- Evidence of flooding, drought, or standing water
- Indicators of changes in soil **hydrology** (lowering or raising the water table)
- Soil conditions and factors affecting the trees, such as frequency of saturation; compaction; erosion; textural gradients; restrictions to root growth from shallow, impermeable layers; and restrictions by roads, rock, urban infrastructure, or building foundations

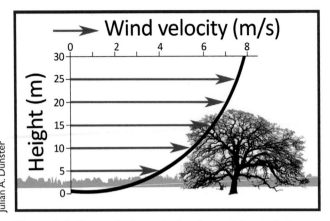

Figure 4.3 Generally, the velocity of a sustained wind increases with height.

Normal storm events may be expected to occur several times during the defined time frame, and they will affect most or all trees in the area. Normal storms may include thunderstorms, snow, and light accumulations of freezing rain in areas that are subject to those conditions.

Extreme storms occur less frequently within the defined time frame and may impact a smaller portion of the region. These events may include severe thunderstorms, accumulations of freezing rain, and straight-line winds.

Abnormally extreme weather events are rare and unpredictable. Tornadoes, hurricanes/typhoons, and microbursts fall into this category, and we do not usually consider their possibility in tree risk

When wind speed reaches hurricane force (greater than 73 mph or 117 km/h), failure of defect-free trees can be widespread.

Government and private weather stations collect information on wind speed and direction in many locations. Websites provide wind data or summaries of data from most weather stations. The data can be graphed (as a wind rose) to model regional or local wind patterns for easier interpretation. Knowing the typical wind patterns for a site may help you assess adaptive growth patterns of a tree.

When evaluating potential **wind loads** during tree risk assessment, you should use typical weather conditions for the region. Normal weather is a meteorological term used to describe the weather based on a location's average temperature, wind, and precipitation for the previous the 30 years.

Normal and extreme wind and storms are less clearly defined. For purposes of tree risk assessment, we consider three types of wind and storm events: normal, extreme, and abnormally extreme. Currently there are no universal criteria that define normal and extreme wind velocities. However, all areas have weather station data and building standards that consider wind load. In addition, some areas have tower crane regulations that specify when they can operate. These data can provide guidance for tree risk assessors.

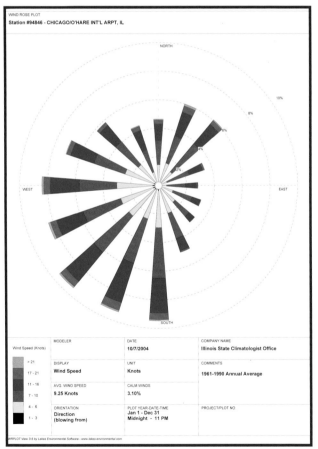

Figure 4.4 Understanding typical wind patterns on a site may help you locate adaptive growth patterns. Many government authorities will provide wind data from specific weather stations. Software is also available to track and map wind direction and speeds (wind rose graphs) and can be used to model regional or local wind patterns.

The Beaufort Scale

The Beaufort scale is a commonly used means of categorizing wind speed on a scale of 1 to 12.

- Force 9 is a strong gale that has a wind velocity of 47 to 54 mph (75–87 kph, 21–24 m/s); this is normal in many areas. Tree failures at or below this level are often associated with defects or conditions that affect a tree's strength or stability.
- Force 10 is a storm wind with a velocity of 55 to 63 mph (88–101 kph, 25–28 m/s); in many areas these wind speeds are considered extreme.
- Force 11 (64–72 mph, 102–116 kph, 29–32 m/s), Force 12 hurricane-force wind (73+ mph, 117+ kph, 33+ m/s), or higher winds are considered abnormally extreme in most geographic areas. At these wind velocities, failure of defect-free trees can be widespread.

Force	Wind MPH	Wind KPH	Classification	Effects on Land
0	<1	<1	Calm	Smoke rises vertically.
1	1–3	1–5	Light air	Smoke drift indicates wind direction, vanes do not move.
2	4–7	6–11	Light breeze	Wind felt on face, leaves rustle, vanes begin to move.
3	8–12	12–19	Gentle breeze	Leaves and small twigs move constantly, light flags extend.
4	13–18	20–28	Moderate breeze	Dust, leaves, and loose paper lifted up, small tree branches move.
5	19–24	29–38	Fresh breeze	Small trees begin to sway.
6	25–31	39–50	Strong breeze	Large tree branches move, whistling heard in wires.
7	32–38	51–61	Near gale	Whole trees move, resistance felt walking against wind.
8	39–46	62–74	Gale	Twigs and small branches break off trees, generally impedes walking.
9	47–54	75–87	Strong gale	Slight structural damage occurs, shingles blow from roofs.
10	55–63	88–101	Storm	Trees broken or uprooted, considerable structural damage.
11	64–72	102–116	Violent storm	Widespread damage to vegetation and structures.
12	73+	117+	Hurricane	Widespread damage/destruction of vegetation and structures.

Adapted from the National Oceanic and Atmospheric Administration (www.spc.noaa.gov/faq/tornado/beaufort.html)

assessments. Thick accumulations of freezing rain or heavy snow, which are rare and unpredictable in most areas, are also abnormally extreme and are not considered for most tree risk assessments.

It is important for tree risk assessors to be familiar with the weather conditions for the region in which they are assessing trees. Tree failures are often associated with forces and additional loads that are associated with the weather. Weather conditions are an important part of the categorization of likelihood of failure.

Wind is the additional load factor most frequently associated with tree failure. For trees, a couple of key points should be kept in mind: (1) **Wind velocity** increases with height. Tall trees and trees atop hills or mountains will experience stronger wind forces than shorter trees and trees at lower altitudes, and (2) Wind flows over and around natural and artificial physical features, and their interacting slopes and surfaces affect wind speed and direction. This variability makes it difficult to accurately model wind close to the ground where trees are affected.

Open water, frozen lakes, roads, or open park areas create less friction (known as the roughness coefficient), so wind velocity will increase in these locations. Look for recent changes that can affect wind patterns or speeds. The shape of the terrain is a factor, too. Wind blowing onto a slope is forced upward, which increases its velocity. If more site-specific information is required, it may be necessary to retain a meteorologist with a specialization in microscale events to make a general prediction of the effects of wind on a site.

In urban areas, wind speeds vary, depending on wind direction and location of structures. Trees on the leeward side of taller buildings, for instance, are exposed to less wind than those growing in the open. In addition, tall buildings and other infrastructure will tend to alter the flow of wind through the spaces. Trees at the corners of buildings can be exposed to the highest wind velocities. Buildings may create wind tunnels, and localized wind patterns with high velocities may result. Development of an asymmetric tree crown is visual evidence of continual wind direction or wind tunneling. Take note of changes such as the demolition of nearby buildings, which can change wind loading of trees.

The manner in which the wind moves around the tree itself will depend on several factors, including the height of the tree, and the shape, extent, and openness (porosity) of the crown. Trees with foliage are less porous and have more **drag** (air resistance), while trees without foliage are more porous and have less drag. Other factors that affect wind movement through a tree are the stiffness of the wood and the degree of protection or exposure it has relative to other trees and nearby structures.

Figure 4.5 The accumulated weight of snow or ice can greatly increase the load on a branch and can cause failure.

Precipitation

Another important weather factor to consider is **precipitation** in the form of rain, snow, and ice. Water is heavy; a U.S. gallon weighs more than 8 pounds (1 L of water weighs 1 kg [2.2 pounds]). Rain can add 25% or more to the weight of a branch with leaves attached, while ice can add at least eight times the weight of a branch, and even more if the tree is still in leaf. The accumulated weight of precipitation distributed over a tree can sometimes be enough to cause limb failure. If the accumulation occurs in conjunction with wind, the likelihood of failure will increase further. Ice storms can leave most of the trees in an affected area damaged, but these storms are unpredictable.

Precipitation also affects tree root failure. Trees are more susceptible to **windthrow** when the soil is saturated, particularly if they are shallow-rooted and the soil is not deep, or there is a high water table. In areas with predictably high precipitation rates or where soil

Module 4 – Site Assessment

Figure 4.6 Trees are more susceptible to windthrow when the soil is saturated.

conditions are a concern, you should attempt to take potential precipitation-related failures into account during risk assessment.

Tree Exposure

The degree to which an individual tree or a group of trees is exposed to the wind directly affects the ability to withstand strong winds. The main categories of exposure are:

- Protected (least exposure to wind)
- Partially exposed (some wind exposure)
- Fully exposed (maximum exposure to wind)

Trees located in the middle of a group of trees, in the interior part of a forest, or surrounded by buildings are examples of protected trees. Trees along the edge of a group are partially exposed to winds from the unprotected side. Trees that are not adjacent to other trees or tall buildings are considered to be fully exposed.

There may be times when you are called upon to perform tree risk assessments in wooded or formerly wooded areas. Trees growing in wooded areas are affected by the surrounding trees in several ways, including light availability, **wind exposure**, and root system overlap. Trees that have grown up in the interior part of a forest will have stability properties that differ from the trees along the edge and from open-grown trees. Interior forest trees experience support and reduced loading from the adjacent trees. Where the individual crowns overlap, the extent of swaying in the wind will be limited. Since trees use resources to increase strength only when necessary, these interior trees will have limited capacity to tolerate increased loading until response growth has developed.

Dominant trees in **forest stands** are more likely to have been subjected to stronger and more variable winds. They will most likely have a fuller crown and stronger root system than understory trees. **Codominant** trees are protected from the wind, and their movement is limited by adjacent trees. Their root systems may be less well developed than those of dominant or open-grown trees, such that removing one or more may destabilize the others. **Subdominant** understory trees will have less well-developed crowns and weaker root systems. **Suppressed** trees will be biologically and mechanically weak; they may fall down easily if they are suddenly exposed.

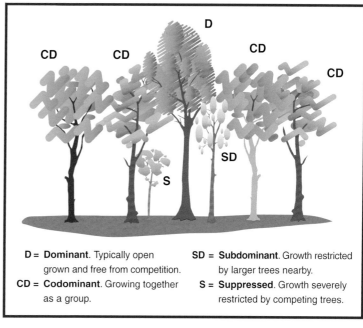

Figure 4.7 Trees growing in wooded areas are affected by the surrounding trees in several ways, including light availability, wind exposure, and root system overlap. This graphic illustrates the relationship of dominant, codominant, subdominant, and suppressed trees in a forest canopy. Changes to the wooded area can affect loading.

Copyright ©2017 International Society of Arboriculture. All rights reserved.

Figure 4.8 Trees that have grown up in the interior part of a forest will have stability properties that differ from the trees along the edge and from open-grown trees. Interior trees that have been abruptly exposed (e.g., from construction, storm damage, or land clearing) may have a higher likelihood of failure.

Root systems develop in response to their environment, including the mechanical stresses they experience. Open-grown trees that are exposed to wind from all directions will normally have well-balanced root systems. By contrast, trees exposed to strong wind from a single direction (prevailing wind) may have asymmetrical crowns and root systems. This may be seen as a network of long, rope-like tension roots on the windward side, and sturdy buttress-like roots on the leeward side. If these roots have been damaged, severed, or decayed, the likelihood of root failure will be increased.

When doing a site assessment, determine whether trees that are fully exposed were always fully exposed. If they recently became exposed as a result of clearing neighboring trees or removing buildings, then failure potential may well be higher. Although **edge trees** may be well adapted to the prevailing winds as a result of continuous exposure, **interior trees** that have been abruptly exposed (for example, from construction, storm damage, or land clearing) may have a higher likelihood of failure.

Well-established edge trees tend to have two obvious characteristics: (1) the branches will mainly be on the open side with fewer on the side toward adjacent trees, and (2) they may be leaning outward as a result of phototropic growth. In some cases, the lean may be greater due to the one-sided crown. In assessing leaning edge trees, check to see if they are partially uprooted.

Topography

A site's **topography**—physical features and changes in elevation—can affect characteristics such as wind load and soil moisture, which can in turn affect the stability of trees. Trees growing on unstable soils and slopes will fail when the slopes fail—failure caused by the soil and topography. Steeper slopes are potentially less stable than gentle slopes and may have thinner layers of soil that are more susceptible to mass movement of the top soil layers (slope creep and landslides). Slopes that are creeping (moving downhill due to gravity) can sometimes be detected by looking for a characteristic J shape at the base of the tree trunks. Usually, the soil at the top of the slope will be well-drained, while soil at the bottom of the slope will be wetter, possibly waterlogged, as water from the upper slope drains and accumulates at the bottom. You can often determine the depth and penetrability by using a common soil probe.

The aspect (compass direction) affects the slope profile and the soil hydrology. Typically, in the Northern Hemisphere, a north-facing slope will be colder and wetter because it receives less direct sunlight. North- and east-facing slopes will usually have thinner soils and will be much rockier. South- and west-facing slopes are usually hotter and drier and have deeper soil profiles. They receive more direct sunlight all year long. The soil and rock weathering processes are much more pronounced on south and west exposures,

Module 4 – Site Assessment

Figure 4.9 A soil probe can be used to determine soil depth and penetrability. Here, the arborist is determining soil conditions under the pavement.

bigger, and, by extension, what factors will adversely affect continued growth and stability. The soil factors to consider are:

- Volume
- Depth
- Moisture
- Compaction
- Quality (texture, fertility, and pH)

The amount of volume needed to sustain and support a tree depends on tree species, size, soil characteristics, and climate (for example, water demand, extreme high and low temperatures). Small soil volumes will limit tree growth, especially the development of strong, structural roots. This is especially an issue in urban areas where space is at a premium, and poor root systems and root cutting are common. An underdeveloped root system may not only be structurally weak, but it will also greatly limit tree growth and health, which diminishes the tree's ability to adapt to the loads it experiences.

Adequate **soil depth** is another important factor in the development of roots. Most tree roots develop in the top 20 inches (50 cm) or less of the soil profile. In ideal soil conditions, many species have a plate of roots growing outward uniformly. Others typically produce oblique roots angling off in all directions; this is sometimes called a heart root system. Few tree species have a single tap root after several years of age.

and the vegetation growing on these areas will be very different from that growing on north or east exposures. The reverse is the case in the Southern Hemisphere. These differences in soil and vegetation can impact root depth and development, which can, in turn, affect tree stability. Trees with deeper, more extensive root systems tend to be more stable.

Soil Influence on Root Development

One of the key site factors that affects tree growth and stability is soil. A well-developed root system will have grown and adapted to support the weight of the tree and the forces it normally experiences. It is difficult to inspect root systems. In a visual assessment, we make some assumptions about root development based on what we see at the base of the tree and the site conditions. We know that a tree that is healthy and growing vigorously is more able to develop response growth to maintain good structure and stability over time. Because root health is important, we look for indications of how site conditions affect root development.

It is reasonable to assume that if a young tree has developed into a large tree, there is adequate soil volume (as well as water, air, and minerals) to sustain the tree. The next question to ask is whether the soil can continue to support trees of this size or

Figure 4.10 Shallow root systems are associated with saturated soils or over-irrigated landscapes.

Figure 4.11 Root health is important, so look for indications of how site conditions affect root development. Surface soil compaction is common where vehicle or pedestrian traffic occurs.

In general, trees with a relatively shallow root system are more prone to failure than those with a deeper root system. Soils may be shallow due to underlying bedrock or urban infrastructure. Shallow root systems are also associated with saturated soils or over-irrigated landscapes. Deeper roots tend to develop on well-drained soils. If the rooting depth is shallow, a wider, more extensive root system is needed to support the tree. You cannot see the roots, of course, so it is difficult to make specific judgments about stability, but understanding soil properties will help you discern possible root issues.

Root development is also affected by the type of soil, its texture, and drainage. Soil texture affects soil moisture: coarse-textured soils will drain quickly, while fine-textured soils with high silt or clay content will drain more slowly. Changes in the soil moisture levels, especially percolation rates, directly affect root development. Any such problem will be made worse by **soil compaction**.

Surface soil compaction is common where vehicle or pedestrian traffic occurs. Compacted soil has reduced space between the soil particles, and, as a result, there is less room for gas exchange or moisture retention. In addition, roots within a compacted soil may be physically crushed or damaged, leading to weakness, decay, or death.

Soils can be assessed through probing, laboratory analysis, moisture testing, and sometimes by reviewing site or development plans. Soil testing and/or analyses are not typically performed in tree risk assessments but may be called for in situations where windthrow or root failure is a major concern. Your understanding of tree biology and experience with risk assessments should guide your judgment of whether to probe, sample, or test. Even when soil testing is not justified, you should consider the belowground component of the site.

Site Disturbance

Any site activity that alters the conditions to which trees have adapted has the potential to affect tree stability. In areas where new development is underway or being proposed, one or more factors may be changed. Changes to consider include:

- Altered wind patterns and velocity due to structures funneling or blocking wind
- Altered levels of sunlight exposure due to new structures and removal or planting of adjacent vegetation
- Changed soil moisture regimes due to compaction, altered natural waterways, installed subsoil drains, and grade changes
- Damaged root systems resulting from compaction, grading, excavation for basements and footings, and installation of underground utilities, services, landscape lighting, and irrigation systems

Figure 4.12 Activities such as moving heavy equipment across the roots, raising or lowering the existing soil grades, and trenching to install utilities or irrigation systems can damage roots and increase the likelihood of failure.

New landscapes installed around existing trees often damage the tree, although the damage may be hidden. Activities such as driving heavy equipment across tree roots, raising or lowering the existing soil grades, trenching to install utilities or irrigation systems, and using irrigation can damage roots and increase the probability of failure.

If roots have been damaged or severed, the health and stability of the tree can be affected. The amount of root loss that is significant enough to affect stability varies with tree species, maturity, crown size, health, site conditions, and climate. As a very general guide, when a third or more of a tree's buttress roots are missing or significantly decayed, stability can be significantly reduced. If one or more large structural roots have been severed within a distance of three times the trunk diameter, you should look for new root growth or root decay to assess whether the tree has sufficient support. Root loss that occurs beyond that distance may be a stability concern if multiple large roots have been severed or destroyed.

Keep in mind that root loss not associated with stability concerns can still decrease tree vitality, which can diminish growth and defense mechanisms.

Although raising the soil grade may damage the existing roots, if the soil is reasonably well-drained, the tree may survive by developing a second root system closer to the surface. Soil against the trunk can allow disease or decay to enter the tree. Some species will respond to fill soil against the trunk by developing adventitious roots. Typically, these adventitious root systems can keep the tree alive, but they might not contribute to structural stability. As the original and damaged root systems decay, the tree may lack sufficient structural roots to resist normal wind and weight forces. You should check carefully for evidence of grade changes near existing trees.

Fill soil, excessive mulch, and vines or other ground cover growing at the base of a tree can obstruct your view of the lower trunk and buttress roots. When you encounter these conditions, you can recommend

Figure 4.13 Any site activity that alters normal site conditions to which trees have adapted has the potential to affect tree stability.

Tree Risk Assessment Manual

Table 4.2 Examples of site factors influencing likelihood of tree failure.

Site Factors	Increased likelihood of failure is associated with	Likelihood of failure is decreased or unaffected by
Wind speed	Unusual winds exceeding local normal maximum (e.g., hurricane, tornado, microburst)	Winds not exceeding local normal maximum
Soil moisture	Excessively wet or waterlogged soils	Dry soils
Soil depth	Shallow soil, especially if prone to waterlogging	Deep soil
Topography	Top of ridge, newly exposed	Valley, low wind exposure, moderate water
Exposure	Forest stand tree recently exposed	Open-grown tree adapted to exposure

CASE STUDY

Bowed tree failing after adjacent codominant trees were removed

stumps

Assignment: Risk assessment on a changed site.

Targets and Site: Established residential area. Three trees at one end of a row of mature trees had been removed after storm damage. The target of concern was the house on the adjacent lot.

Conditions: The bowed Lawson cypress *(Chamaecyparis lawsoniana)*, which was formerly protected by nearby trees (forest canopy codominant), was leaning over the adjacent house. The tree had little taper and a relatively low live crown ratio.

Analysis: Removal of the three trees at the end of the row destabilized the leaning tree. Likelihood of failure was rated as *probable*, and likelihood of impacting the house was considered *high*. Accordingly, the likelihood of failure and impact was *likely*. Consequences of failure were categorized as *significant*, yielding a risk rating of *high*. Adjacent trees could have similar problems in the future because of their new exposure.

Recommendation: Remove the bowed tree and monitor the adjacent trees for increased risk resulting from the new exposure.

Provided by Julian A. Dunster

removal of the plants or excavation of the fill or mulch. Without the view of this important part of the tree, you may not be able to complete the tree assessment.

Changed moisture regimes often create conditions that trees cannot overcome. Trees develop in response to the available soil moisture. If soil moisture changes as a result of construction activities, then the tree's root system has to adapt. Given enough time, and assuming the tree has enough energy reserves available to last while adaptive growth takes place, survival may be feasible. It is not uncommon to see trees go into a slow decline as they struggle to maintain current foliage while they use up energy reserves in an attempt to grow new roots that are seeking better moisture conditions. Examine the topography for signs of possible disruption of water movement.

Summary

Before you start the visual assessment of a tree, it is important to assess the landscape setting. The site assessment will provide you with site-specific factors affecting an individual tree or a group of trees. It will also help you better understand site use patterns, and changes that may have taken place in the past, that are currently underway, or that are planned for the future. Any of these factors can affect the likelihood of failure, the consequences of failure, or both.

Key Concepts

1. Certain aspects of the site may increase or decrease the likelihood of a tree or tree part to fail and may affect the likelihood of that failure impacting a target.

2. Growth and development of a tree are, in part, a response to the forces and conditions the tree has experienced throughout its life. Trees adapt to the site conditions.

3. Wind flows over and around natural and artificial physical features, and their interacting slopes and surfaces affect wind speed and direction.

4. The accumulated weight of precipitation (especially ice) distributed over a tree can sometimes be enough to cause branch or leader failure. If the accumulations occur in conjunction with wind, the likelihood of failure will increase further.

5. Trees are more susceptible to windthrow when the soil is saturated, particularly if they are shallow rooted or the soils are shallow.

6. Interior forest trees that have become abruptly exposed may have limited capacity to tolerate increased loading until compensatory growth has taken place.

7. Short- and long-term issues affecting likelihood of failure associated with site changes are dependent on the capacity of the tree to adapt to the new site conditions. Tree species, health, age, and structural condition impact this capacity.

Tree Biology and Mechanics

– *Module 5* –

Tree Biology and Mechanics

Module 5 Part 1: Wood Structure

Learning Objectives

- Describe the key components of wood structure, including cellulose and lignin, and discuss how each functions in tree stability.
- Explain how trees in good health respond to the loads they experience by producing new wood that varies in amount, placement, and characteristics.
- Explain why an overall assessment of tree health is an important component of risk assessment.

Key Terms

angiosperm	fiber	ram's horn	tracheid
annual rings	flexure wood	rays	vascular cambium
buttress root	gymnosperm	reaction wood	vessel
callus	latewood	response growth	woundwood
cambium	lignin	retrenchment	xylem
cellulose	mechanical stress	secondary xylem	
compression wood	parenchyma	strain	
earlywood	phloem	tension wood	

Module 5 – Tree Biology and Mechanics Part 1: Wood Structure

Introduction

After you have considered the targets and the site, it is time to take a closer look at the tree. When looking at the tree, you rely on your knowledge of how trees grow and respond to loads, coupled with your experiences and observations. This module discusses the fundamentals of tree structure and health, decay, and mechanics that are important for arborists to know prior to undertaking a tree risk assessment.

As a tree risk assessor, you should have a fundamental understanding of tree biology. The purpose of this module is not to provide a comprehensive text on the topic but rather to review how the anatomy and physiology of a tree affect the likelihood of failure. A tree's structure and response mechanisms are part of a complex interaction between the tree and its environment, including the loads experienced and the effects of wood decay. Understanding wood structure is a prerequisite for understanding decay and response growth.

Wood Structure

The way in which wood develops has a direct influence on how trees react to external forces, decay, and other stresses. Every year, a tree dedicates some of its energy to producing new leaves, developing new wood, and growing in all directions. The basic water-conducting system in the wood of a tree consists of a series of cells joined together like pipes in the trunk, roots, and branches. In gymnosperms (mainly conifers), these pipes consist primarily of tracheids, while angiosperms also have vessels.

Vessels and **tracheids** are the water-conducting tissue of trees and, combined with **fibers** and living **parenchyma** cells, comprise the xylem. Wood is **secondary xylem** and is the main source of strength in the branches, trunk, and roots.

The key load-bearing components in xylem cells are cellulose and lignin. **Cellulose** "strings" (which are whitish in color) provide flexibility and strength when under tension. Between and around the strings is a matrix of cell wall material composed of lignin, which can be thought of as blocks that surround the strings. **Lignin** (darker brown) provides stiffness and load-bearing capability when under compression. Together, cellulose and lignin make wood strong yet flexible.

Each xylem cell has layers of specialized cell walls, and each layer contributes specific mechanical attributes to the cell. The primary wall consists of a skeleton of cellulose. The secondary wall, formed after cell enlargement is completed, is rigid and provides strength under compression. It is made of cellulose, hemicellulose, and lignin. The highly lignified cell walls of the xylem's vessel elements and tracheids provide the strength that supports the loads exerted on the tree.

Specialized parenchyma cells that develop radially across the xylem and phloem are known as **rays**.

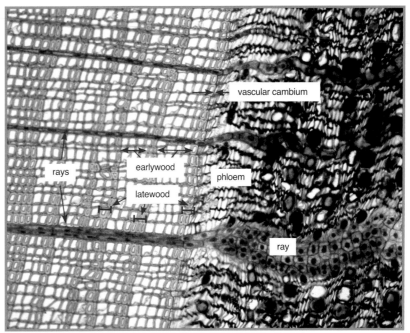

Figure 5.1 Wood is secondary xylem and is the main source of strength in the branches, trunk, and roots. This cross section of a conifer shows xylem on the inside, the vascular cambium, and bark on the outside.

Think of these as similar to spokes on a wheel. Rays transport nutrients and other materials across the short axis of the roots, trunk, and branches to and from the phloem. They also store energy (starch), perform an important mechanical function by binding consecutive **annual rings** together, and help the tree cope with bending stresses. They also make up part of the defense mechanism in fighting the spread of internal decay.

New wood formation occurs annually in temperate climates and nearly continuously in tropical and subtropical areas. In temperate climates, the new wood is normally added as a distinct annual ring, reflecting a clearly defined growing season followed by a period of dormancy. In tropical and subtropical areas, where a period of dormancy may be absent, there can be more than one growth period in one year. The annual ring, typical of trees in temperate climates, often has two distinct wood forms. When the tree first comes out of dormancy, the new wood normally has thin cell walls and larger intercellular space. This is called **earlywood**. Later in the season, the cell walls get thicker and the intercellular space is smaller (latewood). The **latewood** component is the stronger part of each annual ring. These wood properties are affected by rate of growth and by site conditions.

Phloem is not a structural tissue, but it includes the covering that protects the tree—the bark. The phloem serves to transport carbohydrates along branches and down the trunk to other parts of the tree. Both xylem and phloem can change in reaction to physical, physiological, and chemical stresses. The visible changes can be used in assessing likelihood of failure and will be discussed more in Module 6.

Most of this anatomy and physiology applies only to dicotyledonous trees and gymnosperms with **vascular cambium**.

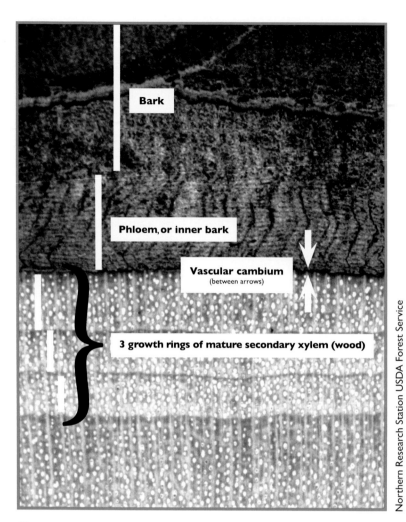

Figure 5.2 In temperate climates, new wood is normally added as a distinct annual ring, reflecting a clearly defined growing season followed by a period of dormancy.

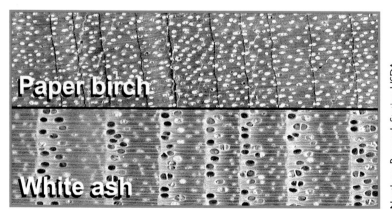

Figure 5.3 The paper birch (*Betula papyrifera*, top) is diffuse porous, and the white ash (*Fraxinus americana*, bottom) is ring porous. Both exhibit earlywood and latewood.

Monocotyledonous trees—such as palms, yuccas, and bamboos—do not have vascular cambium and do not produce annual rings. Instead, they develop bundles of vascular tissue, each one containing xylem and phloem. The strength of the wood is derived mainly from the lignin content in the tracheids and cellulosic fibers within each bundle.

Growth Strategies

Trees go through phases of growth. In their early years, they grow rapidly after germination, moving through the seedling phase into the sapling phase and then the young tree phase. As they age and enter maturity, the rate of growth in height and spread slows, and eventually the overall vigor starts to decline. New wood is still being added, but at a slower rate; annual rings become narrower than when the tree was younger. If the tree can no longer supply adequate carbohydrates, water, and nutrients to sustain growth, old trees may enter a stage of **retrenchment**, dying back in the crown and reducing height and spread while continuing to increase in girth. This retrenchment process adds stability because the reduced crown size decreases the loads. Retrenchment may also increase longevity. Ultimately, trees reach a stage where they are unable to respond and adapt to environmental stresses, and they die. Healthy, young, and vigorous trees have large reserves of stored energy, which allow them to respond to environmental stress, wounds, decay, and major site changes. Unhealthy trees (of any age) have less vigor and reduced ability to respond to environmental stress.

The length of time between germination and death varies by species and within species. Some oaks (*Quercus*) and pines (*Pinus*) may live many hundreds of years, while other species, especially the ecological pioneers in forest ecosystems, may be short lived, perhaps only three or four decades. The lifespan of a tree is an important factor that can influence biology and mechanics. Long-lived species usually have well-adapted biological strategies to overcome the effects of insects, wounds, disease, and decay. Short-lived species often have a weaker response to these factors, relying primarily on new growth.

Tree Health vs. Structural Stability

Do not confuse tree health and tree stability.

High-risk trees can appear healthy when they have a dense, green canopy. This may occur when sufficient vascular transport in sapwood and/or adventitious roots are present to maintain tree health. However, they do not provide adequate strength for structural support.

Trees in poor health may or may not be structurally stable. For example, tree decline due to certain types of root disease is likely to cause the tree to be structurally unstable, while decline due to drought or insect attack may not.

One way that tree health and structure are linked is that healthy trees are more capable of producing new wood to compensate for strength loss associated with structural defects. A healthy tree develops adaptive (response) growth that adds strength to parts weakened by decay, cracks, and wounds.

The structural condition of a tree should be carefully examined and assessed even if the tree appears healthy.

These are two different life strategies: long-lived trees tend to use more of their resources in defending against attack, while short-lived trees tend to use their resources to grow as quickly as possible and to outgrow pests rather than protect against them. Knowing the tree's typical life span and life strategy can be important when assessing tree structure and health. If you know where a tree is in its lifespan and whether it tends to use resources for protection, you can use that information in assessing its likely response to decay or other stress factors.

Response Growth

Tree growth is adaptive and is influenced by loads, the environment, and the availability of essential resources. **Response growth**, a form of adaptation, is the production of new wood in response to damage or additional loads to compensate for higher **strain** (deformation) in the tree's outermost fibers; it includes reaction wood, flexure wood, and woundwood.

Reaction Wood

The two well-documented forms of **reaction wood** are compression wood and tension wood, both formed to counteract gravity and other static loads.

Compression wood, which is common in **gymnosperms** (conifers), is the response to increased static load, mostly due to gravity. It is formed on the downslope side of a stem to make a tree upright, or on the underside of a branch close to the union to support the branch. It is composed of cells that expand longitudinally as they mature. Compression wood typically has smaller, thinner cellulose pipes, with more latewood and high lignin content. By expanding longitudinally, it acts like a hydraulic ram pushing the trunk or limb upward against gravity.

Tension wood is formed in **angiosperms** (hardwoods). Tension wood cells form on the upper side of a branch close to the union, or on the windward or uphill side of the trunk. As they develop, cells in tension wood contract. This contraction acts like a rope pulling the trunk back toward vertical or holding the limb up against gravity. Tension wood has thicker cellulose pipes with less of a lignin matrix between. Angiosperms predominantly produce tension wood; however, some may also respond to static loads by producing compensative growth in the area of compression. Conifers do not form tension wood.

When a branch is cut and viewed in cross section, thicker growth rings are often seen on one side of the branch (lower side for compression wood, upper side for tension wood), especially near where the branch is attached. A similar pattern may be found on the lower trunk, where reaction wood forms on one side to keep the tree trunk vertical.

> ### Response Growth Properties
>
> Properties that show the potential for, or presence of, response growth:
>
> - Crown healthy, vigorous, good color, good growth, and few pests
> - Bark healthy and intact
> - Woundwood well-developed around cuts, cracks, and openings
> - Local increases in wood growth, such as ribs and bulges, near a structural defect
> - Enlargement in diameter of areas weakened by internal decay
> - Distinct demarcations between healthy and damaged tissue
> - Well-developed, wide root flare
> - Corrected trunk lean

Flexure Wood

Some reaction wood development occurs in areas of the tree that are affected by movement caused by changing loads. Sometimes termed **flexure wood**, it is formed in stems and branches in response to wind loading. It is anatomically similar in some respects to tension wood. The development of trunk taper and **buttress roots** is an example of load-responsive (flexure) growth.

Woundwood

Woundwood is a special type of growth that is produced in response to cambial damage. It consists of lignified, differentiated tissue developing from the mass of cells, called **callus**, that forms immediately after wounding. Woundwood is chemically different from, and sometimes denser than, other wood, and resists decay better than normal wood. Its development also reinforces the strength of wounded areas. If a wound or cavity is not readily closed due to the size of the opening or other factors, woundwood may enlarge or curl inward at the edge of the cavity

Module 5 – Tree Biology and Mechanics Part 1: Wood Structure

Figure 5.4 This spruce (*Picea* sp.) was bent by snow in approximately its 13th year. The compression wood (right) developed over several years as the tree righted itself. The difference in color can be attributed to differences in cell wall composition as well as light reflection off the angled tracheids and fibers.

(a feature commonly called a **ram's horn**). The rate of woundwood development is partially dependent on tree health and species characteristics, as well as on **mechanical stress**.

Safety Factor

Trees develop more wood or stronger wood in some areas than is needed to support the tree under normal load conditions—an internal safety factor. The magnitude of the safety factor can vary significantly in different parts of the tree and as a tree ages. This overdevelopment helps to support the tree under extraordinary loads of strong winds, snow, or ice.

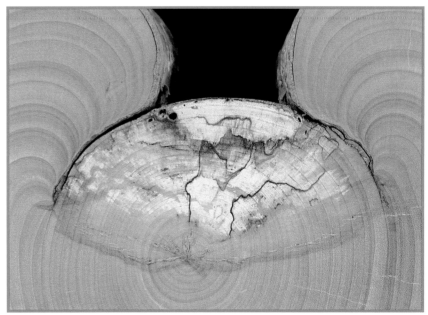

Figure 5.5 Woundwood is developing around this decaying wound and forming ribs. If the tree remains vigorous, the wound may eventually close completely. Note also the zone lines of decay progression and reaction zones.

Tree Risk Assessment Manual

Interpretation of Response Growth

The newest layers of wood experience the greatest torsional, compressive, and tension stress, which is explained in more detail later in this module. As a consequence, the compression or extension of new wood cells under bending loads is greater than the deformation experienced by other cells. The repeated deformation of **cambium** cells influences the number and orientation of new wood cells and their material properties. In general, cells respond to tension and compression by adding varying amounts of lignin or cellulose. New wood that is continually stretched during development will have higher amounts of cellulose present, while new wood primarily under compression will have more lignin.

This uneven thickness of new wood layers provides an external symptom that can be used during risk assessment. You should look for and assess the implications of response growth when evaluating the likelihood of failure. Areas of increased growth should be interpreted as indications of the tree's response to a structural weakness, decay, repeated movement, or other stimuli. It shows both the need for, and the ability to produce, compensative growth due to increased mechanical loads. Keep in mind that the amount of new wood a tree produces depends on species, health, energy reserves, and available resources (water, light, nutrients). The amount and placement of new wood is determined in large part by mechanical stress, but it can also be influenced by physiological and morphological conditions.

When you do identify response growth, try to determine its cause and evaluate its effect on the likelihood of failure. Trees can adapt to weaknesses and stand for many decades if sufficient structural compensation

CASE STUDY

Assignment: Risk assessment for a turkey oak (*Quercus cerris*).

Targets and Site: Formal botanical garden. Targets include parked cars and people walking on the surrounding lawn.

Conditions: There is a pronounced zone of striations down the trunk for about 6 feet (2 m). There is a large cavity about 1.5 feet (0.45 m) to the left of the striations on the other face of the trunk. Overall, the tree appears healthy and vigorous.

Analysis: The pinkish stripes are growth striations, suggesting an area of rapid growth and the formation of response wood. This may be a result of an internal structural weakness developing on one side of the cavity.

Recommendation: Examine the cavity and adjacent wood carefully. Determine the extent of the problem and assess whether response growth is sufficient to compensate. Advanced assessment may be necessary.

Provided by Julian A. Dunster

occurs. Strong response growth properties imply that the tree has substantial stored energy reserves and is working well to maintain biological health and mechanical stability. Poor response growth may suggest that the tree is biologically weak and is unable to fend off decay or resist increased load. It could also indicate insignificant or recent decay. Health and vigor are important factors in predicting the tree's capability to compensate for increased mechanical stress due to decay or increased wind load. Currently, however, there are few guidelines for evaluating the effectiveness of response growth in compensation for structural weaknesses.

Tree Health and Vigor

Part of a basic risk assessment includes a general assessment of tree health and vigor. A tree's ability to resist the spread of decay and produce sufficient wood growth to maintain stability is greatly dependent on general health and vigor. There are several indicators of tree health that you can use to assess tree health and vigor. First, look at overall tree growth and foliage density. Sparse foliage can be an indicator of stress or decline. Dieback of small twigs from the top down is often a symptom of poor root function, possibly due to decay, disease, or cutting. If growth is declining over several years, there may be a problem. Reduced growth can have many causes, but it is always a consideration in risk assessment. Check the foliage, if present. Abnormally small, chlorotic, or necrotic foliage can indicate a tree health issue. Pests and disorders that are unrelated to tree structure may still have relevance if they are contributing to poor tree health or vigor because they may diminish a tree's ability to respond to structural issues.

Figure 5.6 This tree has an abnormally large basal area. Areas of increased growth such as this should be interpreted as indications of the tree's response to a structural weakness, decay, repeated movement, or other stimuli.

Figure 5.7 There are several indicators that you can use to assess tree health and vigor. First, look at overall tree growth and foliage density. Abnormally sparse foliage can be an indicator of stress or decline.

Summary

As a tree risk assessor, you should have a fundamental understanding of tree structure and defense mechanisms so that you can synthesize the data you collect and the observations you make during an assessment, to judge the likelihood of failure. You must understand the fundamental concepts of wood structure to understand tree decay and response growth. Also, because a tree's ability to defend against the spread of decay and to produce sufficient wood growth to maintain stability is greatly dependent on general health and vigor, you must be able to assess tree health.

Key Concepts

1. A tree's structure and response mechanisms are part of an elaborate interaction between the tree and its environment, which includes the loads experienced and the effects of decay.

2. The key load-bearing components in wood cells are cellulose and lignin. Cellulose provides flexibility and strength when under tension. Lignin provides stiffness and compression-bearing capability. Together, cellulose and lignin make wood strong yet flexible.

3. Response growth is the production of new wood in response to damage or loads to compensate for higher movement or load at the cambium; it includes reaction wood, flexure wood, and woundwood. Distribution of new wood is determined in large part by mechanical stress, but it is influenced by tree health, too.

4. A tree's ability to defend against the spread of decay and to produce sufficient wood growth to maintain stability is greatly dependent on general health and vigor.

Tree Biology and Mechanics

Module 5 Part 2: Decay

Learning Objectives

- Describe the CODIT process and how it defends the tree against decay progression.
- Explain how the decay modes of white rots, brown rots, and soft rots differ and the implications of each mode on tree stability.
- List examples of definite and potential indicators of decay.
- Explain how understanding decay modes and the tree parts affected can be used to help assess the likelihood of failure.

Key Terms

barrier zone	compartmentalization	fungi	saprophyte
basal swelling	conk	heartwood	sapwood
bracket	decay	heartwood rot	sapwood rot
brown rot	decomposition	pathogen	soft rot
butt rot	definite indicator	potential indicator	suberin
cavity	discoloration	reaction zone	white rot
CODIT	fungal fruiting structures	rhizomorph	
column of decay		root rot	

Introduction

Wood **decay** is the long-term process of wood degradation by microorganisms. Knowledge of the interactions between tree biology and decay **fungi** is important when assessing likelihood of tree failure. Significant amounts of decay in load-bearing portions of the tree reduce structural strength and increase tree failure potential. Decay fungi may reduce wood strength well in advance of the development of cavities. If most of the cross-sectional area of a tree or tree part is decayed, structural strength will be reduced. However, the mere presence of decay does not indicate that the tree is likely to fail.

Almost all wood decay is caused by fungi, although certain bacteria may play a role. Fungal decay organisms use wood as a food source. There are thousands of tree decay fungi throughout the world, but within a region, a relatively small number are capable of impacting the structural components of living trees. Identification of wood decay organisms can be complex, and positive identification sometimes requires mycological expertise and use of a microscope. Nevertheless, as a tree risk assessor, you should be familiar with the common wood decay fungi in your area and recognize their fruiting bodies such as mushrooms or **conks** (**brackets**). A complicating factor, however, is that fungi are sometimes reclassified or renamed, so the same fungus may have different names depending on the currency of the references.

If you identify a decay organism present, you can search references that describe the mode of decay and, in some cases, describe how aggressive the organism is in that tree species. Some decay fungi are of little consequence to tree stability; others can rapidly lead to failure. Trees defend against the spread of decay, but fungi sometimes overcome the defenses. Knowing the fungus's rate and patterns of decay development can help you assess the likelihood of failure.

Types of Decay Fungi

Decay fungi obtain nutrients by producing enzymes that break down wood components. Most decay fungi attack nonliving cells (**saprophytes**), but some attack living cells (**pathogens**). Virtually all wood decay fungi can survive as saprophytes on dead wood for a period of time. This is important because pathogenic decay organisms can remain viable in decaying logs or in the soil for long periods of time.

Many decay fungi infect trees through wounds. Some are able to attack only damaged **sapwood**; some attack only **heartwood**. Canker rot fungi may attack sapwood and bark at the same time and have the ability to reinvade the wood from the bark. In the absence of mechanical damage and breaching of the bark layers, few pathogens can penetrate the bark and gain entry to the xylem. A notable exception is the *Armillaria* species, which can effectively penetrate the bark of buried root collars and roots.

Figure 5.8 In general, root rots are difficult to detect and assess because their extent and severity cannot be easily observed during a basic inspection. With some fungal decay species, the only definite indication of decay is the presence of fruiting bodies, although their presence may be seasonal, and, in some instances, they may have been cleaned up during routine grounds maintenance.

Because of the differences in decay mode, rate, and consequences, it is important to identify the decay fungus, if possible. You can use field guides, send photographs or physical samples to a plant pathology laboratory, or consult an expert.

Decay fungi are often categorized as white rots, brown rots, and soft rots. Each group affects the mechanical strength of the tree differently because each breaks down different components of the wood. However, like most things in nature, there are exceptions to categorization of each type, and some decay fungi exhibit the characteristics of more than one type of decay. This is further complicated by the fact that fungi are adaptable and sometimes can change their mode of decay, depending on their host species and the conditions present.

White Rots

White rots initially break down lignin, and, to various extents, cellulose and hemicellulose in cell walls, leaving wet, spongy wood that often is white or yellowish in appearance. White rot is classified into two forms:

- Selective delignification occurs when the lignin is decayed ahead of cellulose, leaving behind the cellulose. This generally causes a slow decrease in strength. Delignified wood retains some tensile strength but loses its compressibility.

- Simultaneous white rot decomposes lignin, cellulose, and hemicellulose and leads to a loss of stiffness and tensile strength, severely weakening the wood.

Wood decayed by white rot may have a fibrous appearance, especially in conifers. The white color of the intermediate decay is due to loss of lignin, leaving behind the cellulose, which is whitish in color.

Trees affected by white rot can sometimes adapt to the loss of wood strength by developing new wood around the decayed area. As the wood becomes less stiff, the cells near the decayed area may experience more movement due to loading. This may lead to additional growth and localized areas of swelling (response growth), which may partially or fully compensate for the strength loss in that location. If the wood has been weakened to the point that compression results in fiber buckling, horizontal ridges may form that can be seen in a visual assessment.

White rots are most commonly seen in hardwood trees, though some species also affect conifers. Common root decay white rot fungi include *Armillaria* spp. and *Ganoderma* spp.

Brown Rots

Brown rots primarily break down cellulose and hemicellulose but generally leave the lignin intact. The intermediate decay is brittle and crumbly and is typically brown in color. As the decay advances, the wood shrinks and cracks into blocks, leading to the term "cubical brown rot." Major loss of bending

Figure 5.9 White rots break down lignin, and, to various extents, cellulose and hemicellulose in cell walls, leaving wet, spongy wood that is often white or yellowish in appearance.

strength can occur early in the decay process. Brown rots are often considered to be more serious than white rots due to the greater loss of strength in the decayed wood.

Because brown rots are more rigid and do not flex, response growth tends to be less than in trees with white rot. Thus it is more difficult to use visual indicators to identify trees that may have brown rot.

Brown rots primarily affect conifers, although there are important brown rots in hardwoods, too. Examples of brown rot fungi include *Laetiporus* and *Phaeolus*.

Soft Rots

Soft rots typically decay cellulose first, but they may also modify hemicellulose and lignin, causing cavities within the cell wall. Ironically, the decay is not necessarily softer than brown or white rots. Closer to brown rot than white rot, soft rots are sometimes a combination of bacterial and fungal decay. Soft rots grow more slowly than brown and white rots. Soft rots create pockets of decayed wood that are softer than the surrounding non-decayed areas. The effect is generally localized, less aggressive in rate of spread, and less significant than the brown or white rots.

Soft rots are more commonly found in wood in damp environments. They primarily attack hardwood species. The most important example of a soft rot found in living trees is *Ustulina* (*Kretzschmaria*) *deusta*, which causes pockets of decay and potentially serious structural problems.

Progression of Decay

Being able to estimate the growth rate (spread) of decay within a tree is important. Therefore, if decay is present, you should try to assess not only the extent and location now, but also what you expect

Figure 5.10 Brown rots primarily break down cellulose and hemicellulose but generally leave the lignin intact. The intermediate decay is brittle and crumbly, and it is typically brown in color. As the decay advances, the wood shrinks and cracks into blocks, leading to the term "cubical brown rot."

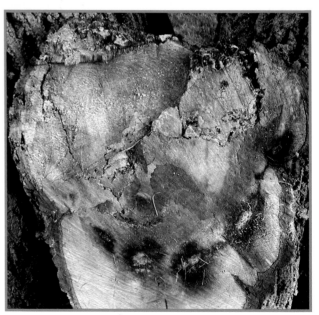

Figure 5.11 Soft rots create pockets of decayed wood that are softer than the surrounding non-decayed areas.

it to be in the future. The key questions to consider are whether the tree's defense mechanisms have contained, and will continue to contain, the spread of decay; whether response growth is sufficient for continued stability; and whether the vitality of the tree indicates continued strong growth and defense in the foreseeable future.

There are no simple rules about rates of decay progression. It is related to:

- Tree and fungus species characteristics
- Ability of the tree to compartmentalize decay
- Age and vigor of the tree at the time of injury
- Type of pathogen (weakly or strongly pathogenic)
- Interaction between fungus and the tree species
- Whether additional wounds have occurred in subsequent years
- Site moisture and temperature

Decay can be described in stages of cell wall **decomposition**. Incipient decay marks the initial stages where the fungus has invaded cells and is becoming established. Cell walls are starting to get thinner and lose some strength. This stage of decay can be detected by laboratory analysis, but usually no obvious visible change has occurred. At the intermediate decay stage, wood **discoloration** becomes visible and wood strength is markedly reduced, but some cell structure is still present. Finally, advanced decay occurs, and all wood strength and cell structure are lost. By the time advanced decay is visible, the wood has already lost its strength.

Compartmentalization of Decay in Trees (CODIT)

Wounded trees do not heal damaged tissues in the way that mammals do. Rather, trees develop cells, tissues, and chemical compounds that contain or isolate the effects of wounds and decay. Changes take place in existing xylem, and new xylem develops around damaged areas as a means of isolating and limiting the progression of decay. In this way, the tree maintains its biological and mechanical integrity.

When the protective bark, phloem, and cambium on a tree are wounded, the outer xylem is exposed, and a process termed **compartmentalization** is triggered. The purpose of compartmentalization is to stop or slow the advance of decay organisms into healthy wood.

Living cells in the xylem around the wound undergo physiological and chemical changes to block fungal growth. The Compartmentalization of Decay in Trees (**CODIT**) model describes a system in which four conceptual "walls," or zones, inside the wood develop to compartmentalize and restrict the spread of decay.

Figure 5.12 This illustration is a graphical interpretation of the four conceptual CODIT walls.

Each wall serves to prevent or slow decay progression in a certain direction. This simple model applies to most—but not all—trees and decays. Although the progression of decay can often be slowed or stopped, some aggressive parasitic fungi are not successfully compartmentalized.

Wall 1 resists the spread of decay up and down the vascular system by plugging tracheids and xylem vessels. Wall 1 is the weakest of the four walls. It is common for separate decay columns, coming from above or below along the axis of a tree stem, to coalesce when Wall 1 fails to stop decay progression.

Wall 2 resists the decay moving radially toward the center of the tree. Composed of the latewood cells in each growth ring and chemicals produced by living cells in this area, it is continuous around the tree, except where the ray cells pass through radially. This is typically the second-weakest wall.

Wall 3 resists decay from spreading from the point of injury around the trunk. It is a series of lateral walls made up of the ray cells on either side of the injury. The living parenchyma cells in the rays create chemical changes that are toxic to decay microorganisms. This process involves production and accumulation of special decay-resistant compounds (phenolic compounds in angiosperms and terpenes in gymnosperms). Because of the pattern and distribution of ray cells in the xylem, Wall 3 is a discontinuous but overlapping wall that establishes a relatively strong defense.

The first three walls are composed of pre-existing wood tissue that undergo changes after injury to serve a defense function (the **reaction zone**). The initiation of physiological changes in wood at Walls 1, 2, and 3 is the first stage of compartmentalization. The second stage is the development of Wall 4.

Wall 4 is also called the **barrier zone**. The new xylem that develops after the injury establishes an interface that prevents decay from moving outward by physically separating the old tissue from the new. Wall 4 is the best defense of the four walls against pathogens and decay, but it is structurally weaker because the new ray cells are not connected to the old ones in the area of the wound. Internal cracks (ring shakes) may develop at Wall 4, spreading around or up and down the tree.

The barrier zone starts as new parenchyma (living cells) that are both anatomically and chemically different from cells produced before wounding.

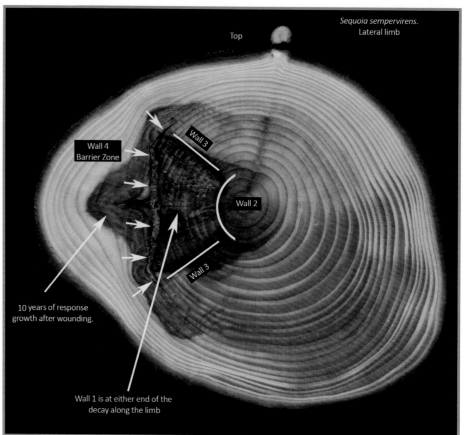

Figure 5.13 Looking at this section of wood, we can see how each CODIT wall limits the spread of decay.

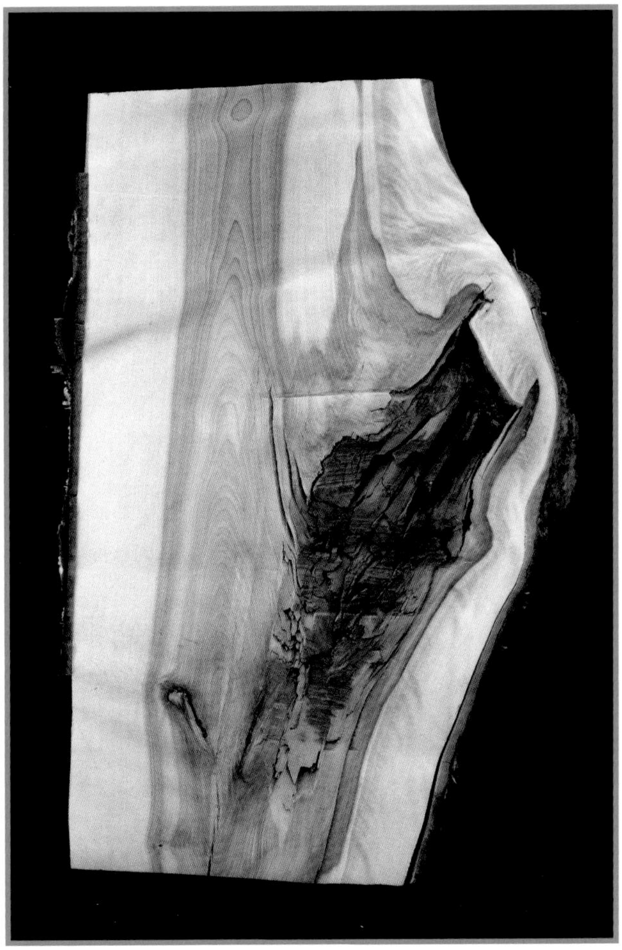

wound. How well a tree compartmentalizes decay is affected by genetics, age, health, vigor, cumulative stress, seasonal timing of injury, virulence of the decay organisms, and the growing environment.

Decay does not spread beyond the wood present at the time of wounding if there is a strong compartmentalization response. The wood that develops in subsequent years will be protected by the barrier zone, and is, therefore, typically free of decay. As a result, the vascular cambium may remain intact and continue to develop outward, away from the decayed areas. The newer, decay-free wood forms a shell of sound wood around the decay column and plays an important role in mechanically supporting the tree. If the barrier zone fails to stop the fungal progression, tree stability will eventually be threatened.

Tree species vary in their ability to compartmentalize decay, and some tree–fungus interaction studies have been published on that topic. Tree failure profiles

Figure 5.14 This tree has completely compartmentalized the decay from an old pruning wound. The barrier zone and the reaction zone are clearly visible. From the exterior, the contained wood is seen as a bump on the trunk.

These new cells contain **suberin**, a waxy substance that protects the cambium from drying. Suberin also inhibits the spread of fungi into the new tissues.

A tree's ability to effectively compartmentalize decay is an important factor in assessing strength and likelihood of failure. It is fairly common for Walls 1, 2, and 3 to fail, allowing decay to spread inside the tree, forming a **column of decay** and, eventually, a **cavity**. Aggressive pathogens can even breach Wall 4, especially when the tree is physiologically stressed. The extent of the wound will also affect the response of the tree. A large wound or a deep injury, beyond the bark and sapwood, will require a larger investment of energy by the tree to contain the decay than would a small or shallow

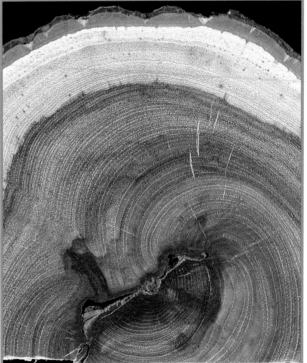

Figure 5.15 As this tree compartmentalized this wound, woundwood ribs formed, and, eventually, the vascular cambium became confluent. New wood completely compartmentalized the wound into a now circular tree.

can be a good source of information on decay spread within species for a given region. There are also a few guidelines that may help you until you become more familiar with the compartmentalization capabilities of species in your region. In general, fast-growing, short-lived deciduous species tend to be poor at compartmentalizing and may decay rapidly, while longer-lived deciduous species, such as many oaks (*Quercus*), are usually much better. Very resinous conifers (such as *Pinus* and *Larix*) are generally strong compartmentalizers and tend to decay more slowly.

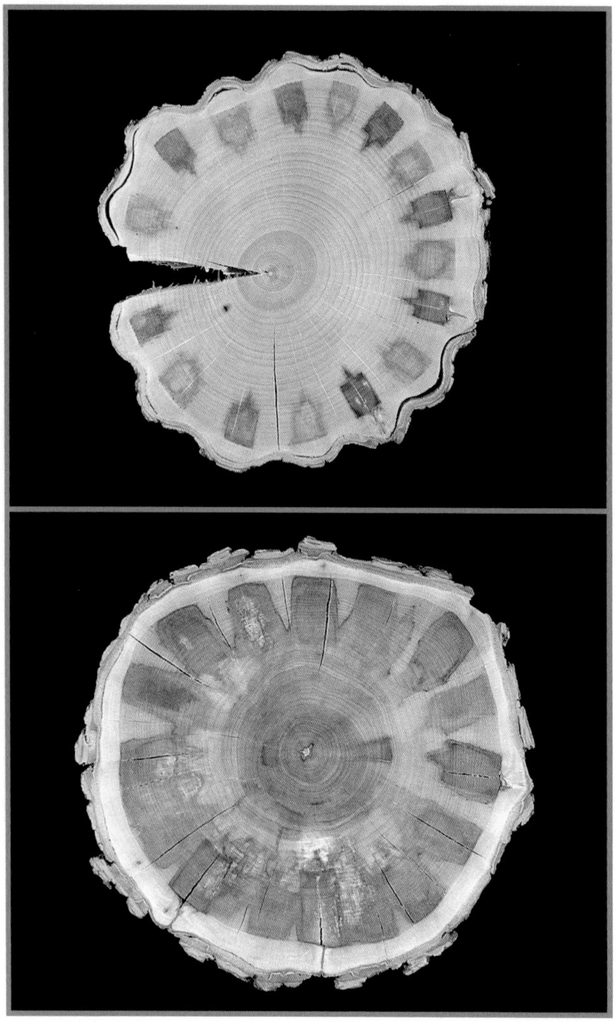

Northern Research Station USDA Forest Service

Figure 5.16 Strong and weak compartmentalization. The tree on the top has effectively compartmentalized the holes made for a research project. Note the response growth that can be seen from the exterior. The tree on the bottom has not been as effective in limiting the spread of decay.

Trees allocate resources for various functions such as growth, defense, and reproduction. Wounding a young, healthy, and vigorous tree will produce different responses than wounding an overmature tree that is already weak. In vigorous trees, there may be enough surplus energy available to create strong compartmentalization responses. The same wound in a low-vigor tree may allow more rapid spread of the decay fungus because the tree may have less available energy to deal with the injury. The allocation of energy toward defense may reduce growth, leading to further loss of energy reserves in later years and, eventually, a more rapid spread of decay.

Location in the Tree

Decay is also classified by its location in the tree, such as root rots, butt rots, heart rots, and sapwood rots. These classifications are not always distinct, however, because some decay may overlap regions or move from one region into another.

Root Rots

Root decay fungi (**root rot**) can infect roots several ways, most often through wounds. A few species have been known to enter as a result of root grafting between infected and uninfected trees, or via specialized structures called **rhizomorphs**. Rhizomorphs look like small-diameter shoestrings or dark-colored roots and are partially responsible for spread of *Armillaria* species. Rhizomorphs can directly infect buried root collars or susceptible roots. Root rots are also spread by waterborne or airborne spores.

In natural environments, most root decay fungi infect through wounds in small-diameter roots, especially those that have been stressed by lack of oxygen due to excessive moisture levels or by lack of water. In the urban environment, damage to larger-diameter roots from trenching, cultivation, or excavation can allow the entry of root decay fungi into these roots. Root decay sometimes progresses from the underside of the roots.

Many root decay fungi will continue growing into the heartwood at the base of the trunk, at which point the decay may be referred to as basal decay or

butt rot. Whenever basal decay is detected in a tree, root decay should also be suspected. However, basal decay may also be associated with decay that entered through wounds in the lower trunk or root collar area.

Because some root rot fungi kill smaller-diameter roots, water absorption and translocation can be affected. This can create water stress, which can eventually be manifested as dieback or thinning in the crown. Crown dieback and thinning, therefore, become useful, but delayed indicators of possible root problems.

In general, root rots are difficult to detect and assess because their extent and severity cannot easily be seen in a basic inspection. Many root rots can directly cause structural root failure or contribute to windthrow. With some fungal decay species, the only definite indication of decay is the presence of fruiting bodies, although their presence may be seasonal, and in some instances they may have been cleaned up during routine ground maintenance. Other definite indicators are basal cavities and areas of depressed bark. **Basal swelling** and fused roots are potential indicators of root decay. When decay affects water uptake, an indication can be crown dieback or other symptoms of loss of vigor (reduced growth, small leaves, yellow leaves, etc.).

Figure 5.17 Root decay fungi can infect roots in several ways, most often through wounds. Many root decay fungi will continue growing into the heartwood at the base of the trunk.

Figure 5.18 This tree has extensive basal rot, but it is also exhibiting significant response growth.

Heartwood Rots

Heartwood rot (heart rot) is the term used to describe decay in the heartwood (center) of the trunk or branch. Although not all tree species have true heartwood, the term is generally applied to decay of the center of any tree stem or branch. Rates of decay in heartwood are variable, based on tree species, condition, and how aggressive the decay fungus is.

When looking at a tree's cross section or an increment core, distinguishing between darkened natural heartwood and discolored wood due to decay can be important. True heartwood is present in many trees and is initiated by aging. It is under strong genetic control, starting in the oldest tissue and developing outward in newer tissue every year. Usually, it is continuous throughout the tree and circular in cross section. By contrast, discolored wood associated with wounding has inner tissue exposed to air, bacteria, and fungi. The discoloration affects the youngest tissue first and proceeds inward. It is seen as one or more columns in the tree's structure and is often irregular in extent across the section and up or down the tree.

Sapwood Rots

Sapwood rot (sap rot) is limited to the sapwood on living or dead conifers and hardwoods, especially when bark and cambium have been damaged by other agents. Loss of sapwood due to decay, disease, or mechanical damage can contribute to tree failure. Loss of outer wood is considered more important to stem strength than is internal decay; outer xylem tissue contributes significantly more than internal fibers to stem strength. A definite indicator of sapwood rot is the presence of numerous, but small, fruiting bodies. Sapwood rot is significant because strength loss can be rapid. Several common sapwood rots are also known to invade healthy wood.

Figure 5.19 Heartwood rot is often irregular in shape and frequently occurs in one or more columns within the tree.

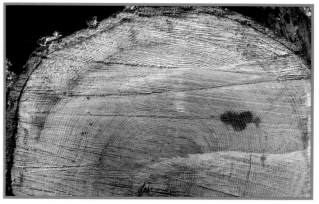

Figure 5.20 Loss of outer wood is considered more important to stem strength than is internal decay. Outer xylem tissue contributes significantly more than internal wood to stem strength.

Figure 5.21 A definite indicator of sapwood rot is the presence of numerous small fruiting bodies.

Indicators of Decay

Decay is often a hidden defect associated with tree failures. Therefore, you need to know how to evaluate the presence and extent of decay and how to consider its significance to the likelihood of tree failure. A tree may appear to be solid and structurally sound, and it may have a dense, green canopy, yet it can have significant decay inside. It is important to recognize common indicators of decay.

Definite indicators of decay occur only when decay is present. The most common definite indicators of decay or internal voids include:

- Cavity openings, nesting holes, bee hives, and other voids or openings to the outside of the tree
- **Fungal fruiting structures**, such as mushrooms, conks, or brackets, that are attached to the tree
- Carpenter ants inhabiting or emerging from defect regions
- Termite emergence from internal nests/tunnels

The difficulty in relying on fruiting structures to determine the presence of decay is that they are not always present or persistent. Some appear for days or weeks when moisture and temperatures conditions favor development and disappear later. In addition, some fruiting bodies are inconspicuous. When they are present and visible, however, their number and location can provide information on the location and extent of decay.

Decay is often present without any visible fruiting bodies, so we must look for other indicators. Cavities are open or closed hollows associated with decay within a tree. They are definite indicators and may indicate more extensive internal decay than can be seen. Cavity-nesting birds, bats, and bees are also definite indicators of decay. Birds will build their nests in hollows or dig out decayed wood to create a hollow. Honeybees often make their hives in tree cavities.

Figure 5.22 *Laetiporus.*

Figure 5.23 *Ganoderma.*

Figure 5.24 *Armillaria.*

Carpenter ants are often associated with decayed wood within the tree. When present, they are mostly inside the tree except when foraging, so it is more common to detect carpenter ants by their pilings at the base of a tree or near the location of decay.

Potential indicators point only to the possibility that decay is present, since other causes for their appearance can be found as well. The most common potential indicators of decay, strength loss, or missing wood include:

- The presence of old wounds or branch stubs that may have allowed decay fungi to enter the tree
- Response growth patterns, such as swelling, bulges, or ridges on a trunk or branch
- Cracks or seams
- Oozing through the bark
- Dead or loosely attached bark, or bark with abnormal patterns or colors
- Sunken areas in the bark
- Termite trails

Figure 5.25 Holes used by wildlife may be small on the outside but can indicate larger cavities within.

Summary

Wood decay is a long-term process of wood degradation by microorganisms. The presence of a large proportion of decayed wood can reduce the structural strength of a tree. Wounding initiates a series of ordered responses summarized by the CODIT model of compartmentalization. You need to know how to evaluate the presence and extent of decay and to consider its significance to the likelihood of tree failure. Because a tree can appear healthy even when there is significant decay inside, it is important to recognize common indicators of decay.

Figure 5.26 Carpenter ants are considered a definite indicator of decay because they nest in decayed wood.

Module 5 – Tree Biology and Mechanics Part 2: Decay

Figure 5.27 Bleeding or oozing from wounds is a potential indicator of decay.

Figure 5.28 Enlarged growth is often from response wood, which is a potential indicator of interior decay.

Key Concepts

1. Almost all tree decay is caused by fungi. There are thousands of tree decay fungi throughout the world, but a relatively small number are important on living trees in any region.

2. White rots break down lignin and, to various extents, cellulose and hemicellulose, in cell walls, leaving spongy, flexible wood that is often white in appearance. As the wood flexes and experiences movement due to loading, response growth usually develops, which may compensate for the strength loss in that location and give visual clues to the presence of the decay.

3. Brown rot decay fungi break down cellulose and hemicellulose but generally leave the lignin intact. Brown rots are stiff and brittle. They are often considered to be more serious than white rots due to the greater loss of strength in the decayed wood before the decay is detected and the lack of response growth to add strength and provide visual indications of the presence of decay.

4. Compartmentalization is a process in which trees limit the spread of decay and maintain the vital functions. The basic components of compartmentalization are the existing annual rings and boundaries between rays, combined with chemical changes and new cell formation.

5. CODIT is a simplified model of the compartmentalization process. The first three CODIT walls are composed of pre-existing wood tissue that changes after injury to serve a defense function (the reaction zone). Wall 4, the barrier zone, consists of new cells that are both anatomically and chemically different from those formed prior to wounding. Wall 4 is the strongest and most important of the defenses.

6. A tree may appear to be solid and structurally sound, and it may have a thick, green canopy, yet it can have decay inside.

Tree Risk Assessment Manual

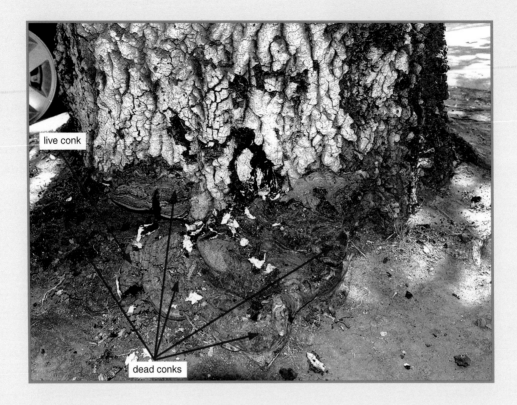

CASE STUDY

Assignment: Risk assessment of street trees.

Targets and Site: Secondary street in a residential area with a row of coast live oaks (*Quercus agrifolia*) growing in a planting strip between the sidewalk and the street. Targets include pedestrians (rare), traffic (occasional), and parked cars (constant). Houses are not within the target zone.

Conditions: At the base of a large, 30-inch (76 cm) diameter tree, several dead and decomposing *Ganoderma applanatum* conks are found on two sides of the tree. A newly dead conk is on a third side, and a small live conk is slightly above the dead conks. No signs of crown dieback are present nor are there any other obvious defects in the tree.

Sounding the base of the tree with a mallet revealed many areas of the entire trunk that sounded dull or possibly hollow. Probing the cavity close to the live conk revealed an area of extensive wood decay, extending at least 20 inches (51 cm) into the trunk. The wood extracted was white, soft, spongy, and very flexible. Careful examination of the bark on the other side close to the area of dead conks revealed similar wood conditions. Decay was extensive, and because it was located at the base, loading was significant.

Analysis: The established time frame was three years. Likelihood of failure was *probable*. Likelihood of impact for the parked cars was *high*, but was *very low* for both pedestrians and traffic. The likelihood of failure and impact for the parked cars was *likely*, and the consequences of failure were *significant* for cars and *severe* for people. The risk was categorized as *high*.

Recommendation: Remove the tree.

Provided by Julian A. Dunster

Tree Biology and Mechanics

Module 5 Part 3: Mechanics

Learning Objectives

- Explain the concept of load and list the primary sources of loads that trees may experience.
- Describe the principle of bending moment, and give an example of how it applies to trees.
- Explain how tension, compression, shear, and torsion affect trees.
- Discuss how trees respond to wind and how they dissipate or transfer wind forces.
- Explain why evaluation of cross-sectional strength of stems is a significant factor in risk assessment.

Key Terms

bending moment	diameter	lever arm	neutral plane	tension
biomechanics	drag	load	radius	torsion
center of force	dynamic	mass damping	shear	
compression	force	mechanics	shell wall	
	gravity	moment	stress	

Tree Risk Assessment Manual

Introduction

We began this module with the statement that trees fail when mechanical stress exceeds strength, and we have examined the biological factors and conditions that affect tree strength. In this section, we will look at the **mechanics** part of **biomechanics** to better understand the loads that trees experience. First, we must understand the terminology that we'll use as well as the fundamental concepts of tree mechanics.

Basic Terms and Concepts

When evaluating tree risk, you need to assess the effects of expected loads on the likelihood of failure. **Load** is a generic term describing the result of various forces acting on a structure. The two most important natural **forces** that exert loads on trees are gravity and wind.

Gravity acts as a constant pull on the mass of the tree, generating load from self-weight, the weight of water (condensation, rain, snow, or ice) on leaves and branches, or even epiphytes. This gravitational dead load can also be increased temporarily during tree care operations with the weight of climbers or equipment.

> A **moment** is a force multiplied by the length of a lever arm to rotate or bend an object.
>
> A **bending moment** is the force that results in bending of an object.
>
> The **lever arm** is the distance between the applied force (or **center of force**) and the point where the object will bend or rotate.

Figure 5.29 Freezing rain can accumulate on twigs and branches, creating a very heavy load that often causes failure.

Energy from wind exerts a **dynamic** (changing) force on leaves and branches from friction and pressure. Wind forces within the canopy vary because the wind energy is transferred to the leaves and branches as the wind moves through the crown, causing branches to bend and twist. In addition, there may be dynamic loads from climbers, rigging operations, or other sources. These forces all result in stresses and strains in the tree structure.

Bending moments are generated by forces acting on a lever arm. Their magnitude depends on the amount of force and the length of the lever arm. One way to understand the significance of this is to consider that the force is magnified by the length of the lever arm. Thus, a tall, isolated tree or a long, lion-tailed branch can experience strong wind loads. The center of wind load can only be estimated because of the varying nature of wind and the tree crown. As a guideline, the center of force is often about 10% above the center of crown; thus, the lever arm distance can be estimated:

Bending Moment (M) = Force (F) × Lever Arm Length (L)

The most common wind-related lateral force on a tree is called **drag**. Because wind force varies with the square of the velocity of the wind, small increases in wind speed can result in a large increase in drag force.

Wind speed usually increases with height above ground, so taller trees and trees at higher altitudes are generally subject to higher winds.

Drag is also dependent on the wind-intercepting surfaces (frontal area and volume of crown facing the wind), wind resistance (drag coefficient) of the tree, and density of the air. The wind-intercepting surfaces and drag coefficient of the tree change with wind velocity due to the streamlining and reconfiguration of leaves, twigs, and branches. The degree of streamlining differs among tree species and with individual trees. Within a range of wind speeds, the higher the wind speed, the more streamlining, until no more is possible.

Loads on a tree lead to internal stresses. **Stress** is a force exerted over an area, mathematically defined using the formula:

Stress = Force/Area

There are four basic stresses within a tree:

- **Compression** is squeezing a material.

- **Tension** is stretching or pulling a material, the opposite of compression.

Figure 5.30 Wind affects the entire tree crown but is often calculated as the force (F) at a single point acting at the center of pressure. The bending moment (M) of the tree is calculated by multiplying the resultant force by the distance from the ground (L) to the center of pressure.

- **Shear** stress occurs when components of a material attempt to slide relative to one another at the interface between tension and compression.

- **Torsion** creates a special type of shear stress caused by a twisting force.

The different types of stresses can occur alone or in combination. When gravity acting on a branch pulls or bends it downward, the bending moment creates tension in the top of the branch (fibers are extended), and the bottom of the branch is under compression (fibers compacted). In the middle of the branch, the **neutral plane** experiences shear stress, where the fibers in tension and compression meet and try to slide in opposite directions. Bending, therefore, involves at least three stresses—compression, tension, and shear—and may also have a torsional component. The same stresses apply to trunks being flexed in winds or leaning over and subject to gravity.

Torsional stress occurs when a branch, trunk, or root twists, leading to maximum stress near the perimeter. Trees with asymmetrical crowns or branches with asymmetrical foliage distribution may experience an uneven wind load, resulting in higher levels of torsion. Although torsion has not received a great deal of attention in tree mechanics, it likely is a significant factor in failures, especially branch failures.

Strength is the ability to withstand stress and strain. Strength is a species-specific property of wood as a material and of the tree as a structure. Breaking stress is the magnitude of stress sufficient to cause failure. In general, wood of temperate tree species is about twice as strong in tension as in compression. For a failure to occur, the stress on the trunk, branch, root system, or associated soil must exceed the strength of the wood fibers or soil.

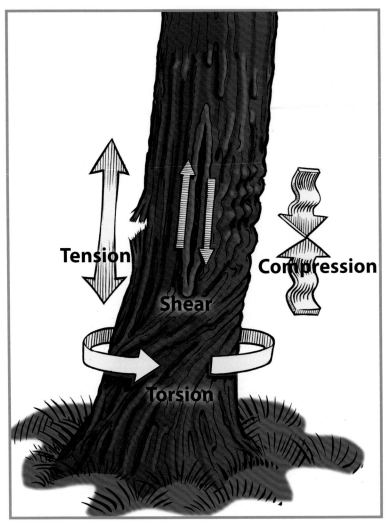

Figure 5.31 The four basic stresses within a tree are compression, tension, shear, and torsion.

Applying Mechanical Principles to Trees

The stress formula shows that stress increases either when there is an increase in load or when there is a decrease in cross-sectional area. Wood cross-sectional area is decreased by loss of wood due to decay. Stress also may be increased locally at a sharp angle bend, canker, or other conditions, all of which are referred to as stress raisers. Under load, a stress raiser may act as a failure initiation point—the point where failure starts. Thus, you should learn to recognize stress raisers and assess how they may be loaded.

Tree risk assessors are rarely capable or equipped to measure actual loads or the actual strength of tree parts. They inspect for and assess conditions that may lead to increased loads or decreased load-carrying capacity as relative factors, as well as conditions that affect a tree's ability to compensate for such changes.

Cross-Sectional Strength

There is a significant body of research investigating cross-sectional strength of trees with internal decay. The central question is: what amount of sound wood is necessary to maintain adequate load-carrying capacity?

The many failure thresholds that have been proposed are based on the ratio of cross-sectional **diameter** of the cavity to the diameter of the tree, typically raised to the third power. This approach is modified from the mechanics formula for a pipe. The pipe formula has been simplified by researchers to focus on stem diameter and the thickness of solid wood (the **shell wall**) that surrounds a centralized cavity. While simplified formulas cannot be used in all cases, the most common threshold allows for the loss of two thirds of the diameter of the stem in the center of a round stem. That is, if the shell wall surrounding an internal cavity is more than one-sixth of the trunk diameter (one-third of the **radius**), then the tree is considered not likely to fail under normal weather conditions.

Though possible to measure and simple to calculate, cross-sectional strength formulas and threshold values have some significant limitations:

- The formulas assume round stems with centrally located cavities/defects, which is often not the case. The location of the decay is important.
- A smaller amount of remaining solid wood may be required for large-diameter trees and those species with stronger than average wood strength.
- The formulas, as presented, do not take into account the size of any cavity openings or the strength of the woundwood that may surround cavity openings.
- There is no consideration of load on the shell wall, so tall trees and short trees, forest trees and city trees are all treated identically.
- There is no consideration for possible increases in the strength of the shell-wall wood added after injury because the tree offsets loss of strength due to decay.

Figure 5.32 Formulas used to estimate cross-sectional strength assume round stems with centrally located cavities/defects, which is often not the case.

Figure 5.33 Many trees are not circular in cross section.

Because cross-sectional formulas consider only the diameter (or radius) ratio, it is important to consider other contributing factors such as species, other defects and conditions, tree height, lean, and crown size/shape. Thus, at best, the shell wall thickness threshold should be considered a starting point for consideration. It should not be adhered to when the tree is not essentially circular in cross section, when the defect is not centrally located, or when loads from wind or supported mass will be low.

There is empirical evidence that trees can tolerate extremely large amounts of internal decay and still remain stable. Very large-diameter trees have been known to remain stable for decades with a remaining shell wall of less than 10% of the radius. Each new growth ring added to the cross section increases the tree's stability if decay is contained or proceeds at a slower rate.

We also know that the location of decay is a critical factor in strength loss. The percentage strength loss is not necessarily proportional to the cross-sectional area of a cavity or decay. If the decay is centrally located, the percentage strength loss will be lower than the percentage of cross-sectional area loss. But if the decay is located at the edge of the cross section, the percentage of strength loss can be higher than that of the area loss. Thus, determining the location of the decay and, if practical, the likely direction of load is important.

Because trees grow and develop wood in locations and with characteristics that respond to the loads they experience, they are typically strongest in the direction of the prevailing load. If trees rarely or never experience strong wind from a different direction than the prevailing wind, they could be highly prone to failure if loaded from that direction.

Height/Diameter Ratio

Stress is also a function of tree height or branch length. A taller tree that has the same trunk diameter as a shorter tree has higher stress in the lower trunk, if all other conditions are equal, due to the longer lever arm. Taper is the change in diameter over the length of trunks, branches, and roots. Slenderness and taper are important factors in the distribution of mechanical stress. The degree of slenderness can be calculated by dividing tree height or branch length by the trunk or branch diameter (H/D) at the base. When H/D is large (that is, the tree or branch is very slender for its height/length), the tree may be more prone to failure than when the taper has moderate or high taper with a small H/D ratio. Reported critical H/D values range from 50:1 to 90:1, but there are many variables. The actual stress is also dependent on the trunk's shape, taper, wood strength, crown configuration, and wind exposure. Therefore, H/D ratio is rarely used by itself to classify likelihood of failure.

Trees and Wind

Wind energy needs either to be dissipated within the tree or transferred to the roots and soil. At low wind velocities, leaf and twig movement dissipates most of the wind energy. As wind speed increases, larger and larger branches move. When wind forces are large enough to bend the trunk, a greater amount of energy is transferred down through the tree to the soil via the trunk and root system.

> ### *Responses to Wind*
>
> Trees respond to wind by:
>
> - Streamlining leaves, twigs, and branches (immediately in the present wind and more permanently when exposed to constant or frequent wind)
> - Reducing height and/or increasing diameter
> - Altering root growth (long, ropelike roots on the windward side; stocky, prop-like roots on the leeward side)
> - Developing smaller leaves and shorter internodes
> - Forming thicker branch unions
> - Changing mechanical properties of the wood (forming reaction wood)

Gusts and turbulence generate loads on the tree at different frequencies and intensities, causing a complex dynamic reaction. Wind moves branches in different directions in a seemingly non-coordinated fashion. This uncoordinated movement acts to dissipate wind energy and slow the movement of larger branches in a process called **mass damping**, which results in reductions in trunk loading and oscillation. Branches and twigs with different diameters and lengths add to damping by moving at different speeds with different loads. If a tree lacks interior or lower branches, there is less damping within the crown, so more force is transferred to the trunk. This can lead to higher stress in the trunk or roots.

When trees are close to other trees or structures or have a normal branch development within the crown, energy is also transferred or dissipated when branches collide. Trees may also lose leaves, twigs, and branches in high winds, thus dissipating energy and reducing wind resistance by providing a smaller frontal area. Branch loss may expose remaining branches to additional stress by reducing the amount of damping and branch collisions that occur.

Figure 5.34 Over time, trees conform their shape to be more aerodynamic when there is a strong prevailing wind. This "flag" tree exhibits reduced height, one-sided branching, and streamlining of smaller branches and twigs.

Assessing Loads on Trees

We have established that load is an important factor in tree failure—failure occurs when load exceeds load-carrying capacity. Therefore, it is essential that you consider loading as part of a tree risk assessment. There are sophisticated ways to measure loads on trees and predict through modeling and calculations how those loads will be transferred and dissipated. Few tree risk assessors have either the instruments or knowledge to do so. More practical load-testing methodologies (described in Module 2) can be employed to test tree strength against predicted loads. Such testing may be justified in some tree risk assessments. Most assessments, however, do not involve advanced testing techniques, so you must employ more basic approaches to assess potential loads and their effects on trees.

The first step is to consider the load from the tree itself as well as that expected from wind, ice, snow, rain, or other factors. Then, assess the areas upon which those loads will act. Consider tree crown size and shape, foliage density, and cross-sectional area. Look for stress raisers—cankers, cracks, sharp bends, and other factors that can magnify the stress. Consider the length of lever arms (tree height, branch length), which multiply the force, especially as those lever arms relate to stress raisers. Look for unbalanced or asymmetrical crowns. Branches that extend beyond the edge of the crown may cause the tree or branch to twist, causing it to fail as torsion forces make wood susceptible to failure.

Tree Risk Assessment Manual

CASE STUDY

Assignment: Risk assessment of black peppermint (*Eucalyptus amygdalina*) adjacent to a picnic shelter.

Targets and Site: Targets include the picnic shelter, cars, and people using the area, with frequent occupancy during the warm months. The shelter has been present for at least 20 years. The soil is compacted and dry. Wind direction and speed vary.

Conditions: The tree lost one codominant stem in a previous storm, three years prior. Decay is visible inside the stump of the lost stem, within the remaining stem, and at the base of the tree below both stems. Some obvious response growth has begun to close the wound on the remaining stem, strengthening the stem at the base. Some root damage is visible, including several small girdling roots. The crown is full and dense, but growth has been declining during the last two years.

Analysis: This formerly vigorous tree is showing some signs of decline, probably due to poor root and soil conditions. Response growth is extensive and appears to have already partially compensated for strength loss due to decay. Loading remains a significant concern due to the lean of the remaining stem, the unbalanced crown, and the density of the crown.

The likelihood of impacting the shelter is *high*, although based on frequent occupancy, the likelihood of impacting a person is *medium*. The consequences of impacting the shelter are rated as *minor*, but they would be *severe* for occupants. Likelihood of failure cannot be determined without further testing. The risk to people in the area could range from *low* to *high*, depending on the likelihood of failure.

Recommendation: Further testing is needed to determine the extent of decay and response growth and the overall stability of the tree. If the likelihood of failure is *improbable* or *possible*, measures should be taken to improve root health and to reduce crown weight toward the shelter. Lean and condition should be monitored. If the likelihood of failure is determined to be *probable* or *imminent*, the risk will exceed thresholds, and the tree should be removed.

Provided by Doug Sharp

Figure 5.35 Sharp bends, such as the one near the branch union of this limb, can be points where mechanical stresses are magnified. Stress raisers, such as torsional cracks and ribs, cavities, taper, and old pruning cuts, can also be seen in this tree.

Another consideration is the crown form of the tree. Decurrent trees—those with a spreading branch architecture—tend to dissipate loads through a wide network of branches before the remaining energy is transferred to the trunk and roots. Excurrent trees have a central leader and tend to transfer energy along the stem and into the roots.

Even if you can obtain actual values (for example, wind speeds or cross-sectional area measurements), it is unlikely that you will be able to perform sophisticated calculations or use absolute thresholds to analyze and employ that data. Instead, you will use your basic knowledge of tree mechanics to assess for factors that may either increase or decrease the likelihood of failure. These load factors are combined with other factors, such as defects, response growth, and site conditions, to help you arrive at some assessment of the likelihood of a tree or any of its parts to fail.

Summary

We know that trees fail when load exceeds load-carrying capacity, so it is essential that you consider loading as part of a tree risk assessment. Within a tree, loads lead to stresses of four major types: compression, tension, shear, and torsion. Stresses can occur alone or in combination. When a force is acting on a reduced area, such as a decayed stem or root, stress is increased. Because of significant limitations in our ability to assess loads or strength quantitatively, we usually focus on relative strength loss or increased load. Although trees can tolerate extremely large amounts of internal decay and still remain stable, you need to be able to determine when strength and stability are insufficient. Understanding the interaction between tree structure and loading will help you make those assessments.

Key Concepts

1. The two natural forces that exert loads on trees are gravity and wind. Their force is magnified by the length of the lever arm. Thus, a tall, isolated tree or a long, lion-tailed branch can experience greater stress.

2. There are four basic stresses within a tree: Compression is squeezing a material. Tension is stretching or pulling a material, the opposite of compression. Shear stress results when components of a material attempt to slide relative to one another. Torsion creates a special type of shear stress caused by a twisting force. Stresses can occur alone or in combination.

3. Strength is defined as the ability to withstand stress without failure. In general, wood of temperate tree species is about twice as strong in tension as in compression.

4. Trees can tolerate extremely large amounts of internal decay and still remain stable. Thresholds for strength loss due to decay are just a starting point for consideration and should not be applied when the tree is not essentially circular in cross section, when the decay is not centrally located, or when the loading potential is low.

Tree Inspection and Assessment

– *Module 6* –

Tree Inspection and Assessment

Module 6

Learning Objectives

- List and describe common defects and indicators of structural problems.
- Explain how to assess the implications of defects relative to the likelihood of failure.
- Describe the structural implications of defects and response growth, and explain how they can interact to either increase or decrease the likelihood of failure.

Key Terms

adventitious branch	corrected lean	lean	sudden branch drop
adventitious root	crack	live crown ratio	sweep
bow	defect	oozing	taper
bulge	freeze–thaw crack	overextended branch	time frame
canker	frost crack	rib	tree architecture
codominant stem	girdling root	seam	visual assessment
compression crack	included bark	shear plane crack	

Module 6 – Tree Inspection and Assessment

Introduction

A key part of risk assessment is to categorize the likelihood of failure—of one or more branches, the stem, or the roots. In a basic risk assessment (described in detail in Module 2), you will walk around the tree, systematically inspecting the crown, branches, trunk, buttress area, and visible roots. You should note and record the defects and conditions that would increase or decrease the likelihood of failure. Understanding what to look for and the implications of what you see requires integration of all the material from the previous modules.

Visual assessment includes looking for and determining the significance of defects and structural conditions. The principle is simple: failure occurs when stress exceeds strength. The assessment needs to consider the load-carrying capacity of the tree and its various parts, as well as the loads that the tree is likely to experience under normal conditions. Remember that the external characteristics of trees are a result of the environment in which they grow and the forces they experience, and trees can adapt over time to the loads they experience. There will be a range of issues to consider, including defects, response growth, and conditions that may increase or decrease the likelihood of failure.

When conducting an assessment, keep in mind that there is great variation among trees, decay fungi, defects, and site conditions. Load estimations lack precision. We have no absolute thresholds for the amount and location of solid wood necessary for stability, and models for predicting failure are limited. Because of this, we must recognize that there is significant uncertainty in our assessments, and we must be realistic about the confidence level we attach to them. Nevertheless, the more data we have, and the greater our abilities to interpret that data, the better we can refine our assessments. This module describes common structural defects and conditions, along with response growth patterns that affect likelihood of failure, and provides guidance for categorizing the likelihood of failure.

What Is a Defect?

For the purposes of tree risk assessment, **defects** are defined as injuries, growth patterns, decay, or other conditions that reduce the tree's structural strength. There are many different forms of defects, each one with its own set of implications for tree strength or stability. Some, none, or all of the defects may be important, while others may be interesting but not serious threats to the structural integrity of the tree. There may also be conditions present that could affect the likelihood of failure but that are not necessarily defects (for example, an especially heavy fruit crop that increases the load experienced). There may be many issues to examine in a tree. Compare what you see to what you think a structurally sound tree should look like in this setting.

Figure 6.1 Visual assessment includes looking for and determining the significance of defects and structural conditions. Consider defects, response growth, and conditions that may increase or decrease the likelihood of failure. Individual defects or conditions may not by themselves indicate a serious structural problem, but in combination with other conditions may contribute to failure.

Tree Risk Assessment Manual

Classifying Likelihood of Failure

Before assessing the likelihood of failure, a **time frame** must be specified to put the likelihood rating in context. The time frame is the length of time (for instance the number of years), for which the assessor is deciding whether or not a specific failure is likely to occur. Without a stated time frame, the rating for likelihood of failure is meaningless. The longer the time frame, the less reliable the rating because conditions that affect failure are prone to change over time.

Likelihood of failure is classified based on an evaluation of defects and structural conditions of the tree or its parts, expected loads, site conditions, and weather. Keep in mind that likelihood of failure must have a time frame specified to have meaning. Definitions of the likelihood of failure categories are as follows:

Imminent	Failure has started or is most likely to occur in the near future, even if there is no significant wind or increased load. The imminent category overrides the time frame stated in the scope of work.
Probable	Failure may be expected under normal weather conditions within the specified time frame.
Possible	Failure may be expected in extreme weather conditions, but it is unlikely during normal weather conditions within the specified time frame.
Improbable	The tree or tree part is not likely to fail during normal weather conditions and may not fail in extreme weather conditions within the specified time frame.

Certain structural defects or conditions are more likely to lead to failure than others. Individual defects or conditions may or may not indicate a serious structural problem, but in combination, and under additional loads, they may contribute to failure. On the other hand, through response growth, trees can strengthen weak areas and support loads, thereby reducing the likelihood of failure. A visual assessment includes looking for and determining the significance of each of the structural conditions, individually and in combination, that increase and decrease the likelihood for failure.

Some guidelines for classifying the likelihood of failure based on certain defects and conditions are presented in this module. For the sake of comparison, the general guidance provided is calibrated for a three-year time frame, unless otherwise specified. In specific photos, however, the time frame is varied and the likelihood of failure is narrowed to a single rating. Be mindful that these are general guidelines and are highly dependent on tree species, loads, response growth, the actual time frame being considered, and other factors.

General Considerations

Careful evaluation of the tree crown, the arrangement of branches, and the condition of individual branches can reveal a great deal about the structural condition of the tree. Certain branch arrangement and attachment configurations are associated with incidences of failure.

Sudden Branch Drop

In some climates, tree genera such as *Acer*, *Ailanthus*, *Albizia* (*Paraserianthes*), *Andira*, *Delonix*, *Eucalyptus*, *Fraxinus*, *Khaya*, *Liquidambar*, *Pinus*, *Populus*, *Pterocarpus*, *Quercus*, and *Ulmus* are known to drop branches unexpectedly in calm conditions and high temperatures. This is called **sudden branch drop** (SBD), and it is not well understood. Because failure occurs without wind load, the material properties must change or cracks must propagate for the wood to fail. These changes in wood properties are most likely related to changes in hydration. Crack formation may be related to temperature changes and drying of wood. Failures typically occur a short distance from the branch union, mainly on horizontally growing branches, and usually during late afternoon through evening. At this time, it is not possible to predict failure or mitigate risk due to SBD.

Trees that have experienced branch failure in the past may be more likely to have branches fail in the future. You should look for patterns of weak branch attachments, decay, decline, poor weight distribution, or other defects associated with previous failures. Failure of additional similarly constructed branches is *possible* to *probable*.

The cross-sectional shape of the trunk may offer some clues about structural weaknesses inside. A vertical trunk will be approximately cylindrical or slightly elliptical. If the cross section is an unusual shape, then there may be major internal issues that need to be investigated. In addition to visible defects such as cracks, ribs, bulges, and cavities, atypical bark characteristics may indicate internal structural problems in the bark.

Poorly attached bark, cracked bark, areas of abnormal bark color or pattern, and bleeding or oozing sap through the bark are all potential indicators of defect that need to be investigated. If they have been present for many years, investigate to see if there is mechanical damage and/or decay occurring behind the symptomatic area.

Flat areas at the trunk flare may indicate missing roots, **girdling roots**, or root growth obstructions such as rocks or underground infrastructure. Other root-related problems such as buried buttress roots can affect root decay or response growth, which, in turn, affect tree stability. If roots are severed, decayed, broken, undermined, or restricted, they may provide reduced anchorage. Structural problems with roots are not the only cause of failures belowground. Some soils allow excessive movement of roots, especially when the soil is saturated, which also can affect stability.

Root problems can be very difficult to detect, and corresponding failure can be difficult to predict. Soil and site conditions around the tree can often provide information on changes that have occurred, which may have damaged the root system or affected soil strength.

Branch Attachments and Associated Defects

In general, branch attachments are strongest when the parent branch is at least twice the diameter of the subordinate branch, the branch union is U-shaped, and there is adequate wood to securely hold the branch in place. Codominant stems, unions with included bark, and adventitious branches can represent potential weaknesses, especially when combined with other conditions that can increase the likelihood of failure.

Dead, Broken, and/or Hanging Branches

A lack of live bark, foliage, buds, or leaf growth, and missing, dead, or loose bark are all indicators of dead branches. In some species, dead branches may remain mechanically stable for many years after death. The likelihood of failure rating for dead branches ranges

Figure 6.2 In some species, dead limbs may remain mechanically stable for many years after death. The likelihood of failure rating for dead branches ranges from *possible* to *imminent*, depending upon species, branch weight, type and extent of decay, infestation by wood-consuming insects, and length of time that the branch has been dead. In this case, the assessor rated likelihood of failure of the dead branch *probable* in a one-year time frame.

Figure 6.3 Broken branches are branches that have experienced a structural failure. The likelihood of failure of most hanging branches (hangers) is considered *probable* or *imminent*. This hanger was assessed as *imminent*.

Figure 6.4 Adventitious branches, such as epicormic shoots or watersprouts, are weakly attached in the short term because minimal wood has formed to hold the branch in place. If these branches are attached near a cut or broken branch end, decay developing from the opening may further reduce the strength of the attachment over time.

from *possible* to *imminent* in a three-year time frame, depending on species, branch weight, type and extent of decay, infestation by wood-consuming insects, and length of time that the branch has been dead.

Broken branches are branches that have experienced a structural failure. They may remain partially attached at the point of breakage, or they may have completely detached and started to fall. Branches that are broken and lodged in the crown of the tree are called hangers or lodged branches. The likelihood that the branch will continue its fall depends upon how it is being held and if the branch is decaying. The failure likelihood in a three-year time frame of most hangers is considered *probable* or *imminent*.

Adventitious Branches

Strong branch attachments form when branch and trunk wood develop together over time. **Adventitious branches**—such as epicormic shoots or watersprouts, which often are produced after storm-related branch breakage, lion-tail pruning, or topping—are weaker because, in the short term, less wood has formed to hold the branch in place compared with a normal branch of the same diameter. If these branches are attached near a cut or broken branch end, decay developing from the opening may reduce the strength of the attachment over time. If decay is not present, these branches should be considered *possible* to fail within three years. If decay is present, the likelihood of failure may be considered *probable*, depending on load and weight distribution. If significant new holding wood has developed and no decay is present, the likelihood of failure may be reduced to *possible* or *improbable*.

Codominant Branches

Two stems or branches that are approximately equal in diameter and arise from the same location are called **codominant**. Typically, strong branch attachments develop when the size of the branch is less than one-half the diameter of the parent stem. When the diameters of the branch and parent stem are similar, however, the attachment may be weaker. Stem orientation, weight distribution, and branch configuration will affect stress at the union, making failure more or less likely.

The likelihood of failure is also affected by the shape of the union and by the presence or absence of included bark. Stems that divide in a gentle U-shape tend to be stronger than those with a sharper V-shape. Likelihood of failure of codominant stems with a V-shape within three years is often considered *possible* to *probable*. Failure of U-shaped unions is *improbable* to *possible*.

Included bark is bark that is embedded between a branch and its parent stem or between codominant stems. It decreases the strength of the attachment. Likelihood of failure of branches with included bark is largely dependent on the local stress experienced at the branch union. A crack at the attachment increases the likelihood of failure. The presence of response growth at the union indicates that the union

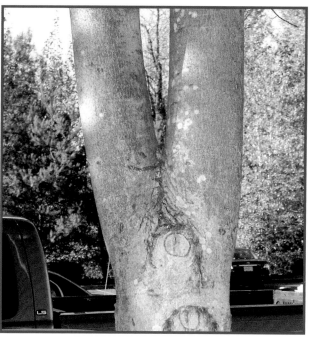

Figure 6.6 Likelihood of failure of a codominant stem with a V-shape, as seen in this tree, is often considered *possible* to *probable*. In this case, the assessor rated it as *possible* within a five-year time frame.

Figure 6.5 Two stems or branches that are approximately equal in diameter and arise from the same location are called codominant. When the diameters of the branch and parent stem are similar, the attachment may be weak. The likelihood of failure of this codominant branch configuration was rated as *possible* for a three-year time frame.

Figure 6.7 Failure of U-shaped unions, as seen in this tree, are *improbable* to *possible*. This was assessed using a five-year time frame and was assigned an *improbable* likelihood of failure.

Tree Risk Assessment Manual

Figure 6.8 Included bark is bark that is embedded between a branch and its parent stem, or between codominant stems. It decreases the strength of the attachment.

be considered *possible* to *probable*, depending upon load, weight distribution, and union shape. If a crack or decay is also present near the point of multiple branch attachments, likelihood of failure is increased.

Overextended Branches

Overextended branches are those that grow beyond the normal crown area. Branches outside of the normal crown may experience higher loads during high winds, freezing rain, or snow storms, and are more likely to fail in these conditions. Horizontal branches tend to be stronger at the union than those growing at an upward angle, when all other conditions are equal.

is under strain (deformation). If there is enough response growth, the likelihood of failure may be reduced. The presence of cracks in the response wood may indicate an increased likelihood of failure.

For many species, failure of codominant stems with included bark may be considered *possible* to *probable* within a three-year time frame. If there is significant decay in or near the union, the likelihood of failure can be *probable* to *imminent*, depending on loads and weight distribution. When considering pruning of one side of a codominant stem, note that there will be no branch collar at the base, so a larger-than-normal column of decay may develop after pruning. This may be a consideration when weighing the trade-offs between removal, installing a support system, or retaining and accepting the associated risk.

Multiple Branches

When several branches originate from the same place on the stem, there may not be enough space for sufficient wood to develop around each branch to provide adequate support. With this type of branch formation, the branches tend to be more weakly attached than single branches of the same size. For some species that tend to fail at this point, the likelihood of branch failure within three years should

Figure 6.9 When several branches originate from the same place on the stem, as in this previously topped tree, there may not be enough space for sufficient wood to develop around each branch to provide adequate support. The likelihood of one of these branches failing in a five-year time frame was rated as *probable*.

Figure 6.10 Branches outside of the normal crown may experience higher loads during strong winds, freezing rain, or snow storms, and are more likely to fail in these conditions. A tree risk assessor rated the likelihood of failure of this overextended branch to be *possible* in a two-year time frame.

Trunk Defects and Conditions

Injuries and Cankers

Past pruning cuts, holes, branch stubs, scars, missing bark, and wounds caused by mechanical, animal, or insect damage are all potential points of entry for decay organisms. Wood at this point may be damaged, structurally weak, or missing entirely. When these features are seen, decay may be present. Sometimes it is impossible to see if decay is present within a tree, but you can look for indicators to know when to investigate further. Vigorous callus around the wound suggests the tree has a strong response and is actively working to contain the decay. Well-defined callus but soft or missing wood within the injury area may indicate advanced areas of decay.

Root decays resulting from injuries are sometimes overlooked because they are not as visible or as easy to locate, and it can take significantly more time to undertake an advanced root assessment. Yet decay in roots may be far more extensive than simple detection procedures would indicate. Decay in the base of the tree may indicate root decay because most fungi that decay the butt or base of the tree start in the roots.

Where injuries are noted, look for signs of decay, sometimes indicated by one or more fungal fruiting structures. Symptoms that decay might be present also include changes in growth rate, changes in foliage color, or dieback of one or more parts of the crown.

Cankers are areas of dead or dying wood, cambium, or bark caused by diseases or repeated mechanical injury. Perennial canker-causing organisms (target cankers) infect the tree through a bark wound. They tend to grow slowly during dormancy and are then surrounded by a new callus roll during the active growing season. Diffuse cankers are shallow and tend to advance more rapidly in the bark, often overwhelming the tree's defensive abilities. The bark at the margins may be discolored.

If there is a lack of new wood growth or significant loss of sapwood, cankers may affect the likelihood of failure. Large canker faces may engulf substantial amounts of the trunk. As a guideline, if more than one-third of the circumference is affected, the probability of failure will be increased, depending upon species, exposure, and whether or not decay is also present. Some disease cankers affect species of all ages and kill the tree, starting in the canopy and spreading down to the trunk.

Bulges

Bulges are pronounced areas of growth whose shape does not match the rest of the tree. Bulges are usually response growth and can occur in response to internal decay, cracks, and wounds. They also can result from localized compression failure where the weight of the tree is simply crushing the trunk downward. This occurs when the trunk is hollow and the remaining wall thickness is thin.

A bulge is caused by new tissue formed as a response to movement. The additional wood that is formed reinforces the wood structure at the weak area. A similar effect is seen where a cavity opens to the outside and new tissue forms around the cavity opening.

Bulges are also common at the root collar and often indicate one of two issues: (1) loss of wood due to root decay moving into the lower trunk may trigger response growth to thicken the remaining shell wall around the decay, or (2) a response to contact stress where the tree has grown against and around an object—a pipe, old railing, or rock, for example.

A particular form of bulge is seen on either side of codominant stems. Once the stems have reached a point where there is no more space to grow inward, new tissue often forms on either side, forming a pronounced rib or bulge spanning the base of the two stems. The bulges are always seen at opposite sides of the two stems. There is usually an area of included bark underneath, and decay may be present. In advanced growth stages, it is common to see cracks in these bulges because the two stems have moved away from each other. This area of additional wood growth is always a high stress area, and failure is very common at this point. In some cases, especially where two or more stems join close to the ground, and are, therefore, often in compression, you may see the side bulges expanding laterally to form a ring of expanded growth around the circumference of the stem. This indicates stress being distributed around the entire tree.

A trunk or branch that has an elliptical shape (ribs or bulges on opposite sides) would most likely have a shear plane crack through the wood, with new growth forming around the cracks. Or it could be the point below two codominant stems where the new wood has grown over the base of the two stems to cover up the crack and included bark underneath.

Occasionally, you may find bulges at the branch attachments, either on the top or the underside. This type of bulge may indicate an internal problem such as decay or cracks. If the bulges are on the horizontal plane, it might be that gravity is overtaking the strength of the attachment point (a horizontal shear plane crack is developing), and new supporting wood is being added. Bulges on the top and bottom could be a vertical crack that developed in an abnormal wind (caused by lateral motion). As with all other bulges, a more detailed assessment may be required to assess the severity of the defect or condition.

Ridges

Response growth and the effects of gravity can result in ridges on curved trunks or branches. They are often more apparent on smooth-barked trees. If there are no signs of cracking or oozing, the ridges probably do not affect the likelihood of failure. If there is cracking or buckling associated with the ridges, or if there is oozing, there may be decay behind the ridges.

Figure 6.11 A pronounced rib or bulge spanning the base of the codominant stems is formed when the stems have reached a point where inward growth is no longer possible (no space to grow), and new tissue is formed on opposite sides of the two stems.

Module 6 – Tree Inspection and Assessment

Figure 6.12 Response growth and the effects of gravity can result in ridges on curved trunks or branches. They often are more apparent on smooth-barked trees. The likelihood of failure for these trees was rated as *improbable* within five years.

Figure 6.13 Severing major, structural roots will decrease the stability of a tree. The likelihood of failure increases with the number of roots cut and decreases with the distance from the trunk. The likelihood of windthrow within three years for this tree was rated as *probable*.

Root Defects and Conditions

Dead, Decayed, or Missing Roots

Dead, decayed, or missing roots can be identified by a cavity in the root collar, a canker that extends to the soil line, or a visibly pruned or broken root stub. Likelihood of failure increases with the percentage of severed or decayed structural roots that have been cut and can be decreased by the presence of response growth.

Abnormal Root Flares and Basal Swelling

Larger-than-normal flare at the base of the tree and fusing together of buttress roots may indicate that root decay is present. Both of these conditions are forms of response growth that may indicate the presence of defects, but they also add strength to resist bending and aid stability.

Visual Indicators of Root/Soil Defects That May Indicate Reduced Stability

- Dead or missing roots
- Fungal fruiting structures
- Lack of root flare
- Stem-girdling roots
- Dead or loose bark
- Wounded roots
- Root cuts
- Soil mounding or cracking
- Crown dieback or decline
- Restricted root space
- Soil erosion
- Excessive soil moisture
- Adventitious roots
- Termite nests/mounds at tree base

Visual Indicators of Compensation

- Wide root flare
- Fused buttress roots
- Large adventitious roots

Buried Root Collar

For most open-grown, mature trees, if the buttress roots are not visible, then it is likely that there is soil against the trunk of the tree. Soil against the trunk may accelerate sapwood or heartwood decay.

Stem Girdling

Where there is a root, strap, chain, rope, rock, concrete, or other object tightly pressed against the lower trunk or buttress roots, growth and response growth can be severely limited by stem girdling, which can lead to both health and stability problems.

Flat Areas on the Trunk

Flat areas at the trunk flare may indicate missing roots, girdling roots, or root growth obstructions such as rocks or underground infrastructure.

Oozing

Oozing from the lower trunk and buttress roots may be associated with root disease. Root disease may impact tree health, such as from a *Phytophthora* infection, or it may affect stability, as with an extensive *Armillaria* infection.

Adventitious Roots

Adventitious roots develop when the trunk is buried or there is a loss in root function. They can indicate root problems and can be compensation for weakness. In some tropical species [for example, Angsana (*Pterocarpus indicus*)], the presence of adventitious roots at the union may indicate a crack and should be investigated further.

Decay

The basics of decay and the tree's reaction to it are discussed in Module 5. In this module, decay is considered a defect of significance that can affect the branches, trunk, and/or roots. The presence of a large proportion of decayed wood or wood that is missing may reduce the structural strength of a tree. However, the mere presence of decay does not always indicate that the tree has been significantly weakened.

Figure 6.14 For most open-grown tree species, if the buttress roots are not visible, then there is soil against the trunk of the tree, which may accelerate sapwood or heartwood decay. The tree was recommended for an advanced assessment of the root crown.

Figure 6.15 Flat areas at the trunk flare may indicate missing roots, girdling roots, or root growth obstructions such as rocks or underground infrastructure. This tree had a very large stem-girdling root, which may have contributed to tree failure.

Decay is often difficult to detect, and, once detected, it is not always easy to assess its significance to likelihood of failure. In many cases, the tree may have been wounded years before your site visit. Some fungi take many years to create structural problems, while others are aggressive and cause major problems within a few years. Many fungal pathogens have characteristic patterns of decay, and to some extent, knowing the pathogen and how it develops can help to determine the likely internal conditions. New wood and bark

Figure 6.16 Larger-than-normal flare at the base of the tree may indicate that root decay is present. This condition is a form of response growth that may indicate the presence of defects, but it also can add strength to resist bending. Failure of this tree within a three-year time frame was rated as *possible*, with an advanced assessment recommended.

Common Definite Indicators of Decay or Internal Voids

- Cavity openings, nesting holes, bee hives, and other voids or openings to the outside of the tree
- Fungal fruiting structures, such as mushrooms, conks, or brackets that are attached to the tree
- Carpenter ants inhabiting or emerging from defect regions
- Termite emergence from internal nests/tunnels

Common Potential Indicators of Decay, Strength Loss, or Missing Wood

- Presence of old wounds or branch stubs that may have allowed decay fungi to enter the tree
- Response growth patterns, such as swelling, bulges, or ridges on a trunk or branch
- Cracks or seams
- Oozing through the bark
- Dead or loosely attached bark, or bark with abnormal patterns or colors
- Sunken areas in the bark
- Termite trails

may have grown over the original wound, leaving little evidence of internal conditions, or there may be decay that entered from root rot, a pruning wound, or at a branch attachment point where there is included bark. In all cases, the extent of decay, its structural implications, and the amount of response growth are the key factors to assess.

Decay usually cannot be seen within a tree, and decay fungi may reduce wood strength well in advance of the appearance of fruiting bodies or the development of internal voids. Look for indicators to help you know when to investigate further. Potential indicators are signs or symptoms that decay may be present, while definite indicators signify that decay is present.

In a basic assessment, the extent of internal decay may be estimated by interpreting external indicators and by using the CODIT model. However, external indicators of decay usually do not provide an accurate estimate of extent. Accuracy in estimating the extent of decay and in determining the location of internal hollows can be improved with techniques such as sounding the tree with a mallet and listening for tonal differences. When you need more information on the extent of decay to assess the likelihood of failure, an advanced assessment may be required.

The likelihood of failure associated with decay depends on the tree species, extent and location of the decay, the presence of other defects, the expected loads, and the amount of response growth.

Cavities and Other Openings

Cavities are seen at sites of past injury that allowed decay entry. It is not unusual to find trunks and branches with cavities, but their presence is not an automatic reason to condemn the tree. Cavities are often used by wildlife as a habitat. The fact that birds or small mammals are using a cavity is an initial indicator that the cavity may be quite extensive. Most cavities in use as habitat tend to be dry inside, which is useful to know. A wet cavity is collecting rainwater either from direct entry into the cavity opening or from seepage coming down through a column of decay above the cavity opening. The latter is more serious because it suggests that the decay column is linked to one or more other openings higher up. In aggregate, these columns of decay may be a serious structural issue, and likelihood of failure may be increased.

The extent of a cavity can be determined by probing. Be careful when probing, especially if you are putting your hand in the cavity. Check first to see that the cavity is not occupied by wildlife such as squirrels, raccoons, snakes, scorpions, or spiders. Sounding the wood around the cavity with a mallet may give some initial sense of the extent of the cavity. In some cases, a more advanced assessment will be required to accurately define the extent of the cavity.

The cause of the cavity, its age, and its condition can provide important information. You need to know if

Factors That _Increase_ the Likelihood Rating of Decayed Branches and Trunks

- Multiple pockets of decay
- Branch that is not vertical, the greater the angle from vertical; the higher the rating
- High end weight
- Lack of internal branching needed for damping
- Tree species known to allow rapid advancement of decay
- Decay fungi known to advance rapidly

Factors That _Decrease_ the Likelihood Rating of Decayed Branches and Trunks

- Significant amounts of response growth
- Small crowns
- Low wind exposure
- Decay-resistant/strong compartmentalizing tree species
- Decay fungi species known to advance slowly

Figure 6.17 Cavities are common at locations of past injury that created a place for decay fungi to enter the tree. Examine cavities to determine their extent, whether the decay is still expanding, and whether the tree is producing response growth. Based on a five-year time frame, the assessor rated the likelihood of failure of the trunk of this tree as _probable_.

there is active decay still taking place, adaptive growth offsetting any strength losses, the overall vitality of the tree, the extent of the cavity up and down the trunk, and its location and structural implications to decide if the cavity is or is not a significant problem.

Cracks

Cracks are a separation of the wood fibers. They are the result of a localized failure caused by excessive compression, tension, or torsion. They also can be a result of frost (freeze–thaw). Some cracks are very significant; others are not. Cracks are indicators of overloading and should be investigated in more detail to see if structural integrity has been compromised.

Cracks can occur along or across roots, trunks, or branches and are seen when they go through the bark. Longitudinal branch cracks that go through the branch or trunk are called shear plane cracks and most commonly occur when wind, rain, freezing rain, or snow overloads the branches or trunks. Cracks also occur when stems or branches are torsionally stressed or below codominant unions that are splitting apart. Cracks can occur in wood with or without decay.

The wood behind cracks may be sound, decayed, or completely missing. A crack can be:

- A surface crack that starts on the bark surface and penetrates into the wood
- A crack starting inside the wood and radiating outward
- A crack that extends all the way through the branch, stem, or root

Figure 6.18 Cracks are a separation of wood fibers. They are the result of a localized failure caused by excessive compression, tension, or torsion. With a one-year time frame, this branch was assigned an *improbable* likelihood of failure.

The typical tree response to a crack is to produce callus followed by woundwood along the margins. Over time this response growth may become inrolled into the opening. The woundwood may eventually meet and form ribs. If the cambium layers from the two sides merge, they can become continuous (confluent) and the crack will be completely enclosed.

When assessing cracks, you should try to determine the following:

- Extent of the crack, both length and depth
- Presence or absence of decay
- Presence or absence of response growth
- Cause of the crack

Cracks that are fresh and associated with decay can greatly increase the likelihood of failure, often to the level of *imminent*. Cracks that penetrate all the way through a branch or stem with no decay will greatly increase the likelihood of failure to *probable* or *imminent*. Cracks that have woundwood surrounding them, including the peak of the cracks, will have a slightly lower likelihood of failure. Cracks that occur across the short axis of the root, trunk, or branch are symptomatic of wood that has advanced decay. Likelihood of failure within three years is *probable* to *imminent*.

Figure 6.19 Cracks can occur when stems or branches are torsionally stressed. This tree was given a *probable* likelihood of failure in a three-year time frame.

Tree Risk Assessment Manual

Figure 6.20 Cracks located along or close to the center of the long axis (vertically in trunks, or along the length of branches) are also known as shear plane cracks, neutral plane failures, or hazard beams. They occur at the transition zone between the opposing forces of tension and compression when the part is loaded. Based on a five-year time frame, the assessor rated the likelihood of failure (and falling) of this branch as *probable*.

Shear Plane Cracks

Cracks located along or close to the center of the long axis (vertically in trunks, or along the length of branches) are known as **shear plane cracks**, neutral plane failures, or hazard beams. They occur at the transition zone between the opposing forces of tension and compression when the part is being bent. If a crack is obvious only on one side of the cross section, it may be a result of torsional forces. If the shear plane crack is visible on both sides, 180 degrees apart, the entire component (branch, trunk, or root) has split. Shear plane cracks can result from bending forces caused by limbs or trunks that are overloaded by gravity, snow, ice, or wind. In branches, they sometimes occur at a curve in the branch.

If the separation does not result in complete failure, there will be two pieces of wood acting independently of each other, and the overall strength of the component (trunk, branch, or root) is decreased. If the forces that created the crack do not recur, the tree may develop response growth, typically seen as ribs growing over the crack from either side. Investigate the crack carefully to determine its extent and internal conditions. In some cases, a shear plane crack will be present in a trunk because there is decay at the base. In that case, you should consider what the decay pathogen is and whether it represents serious root rot problems as well.

Failure of trunks or branches from shear plane cracks is very difficult to predict, and likelihood of failure within three years may range from *possible* to *imminent*.

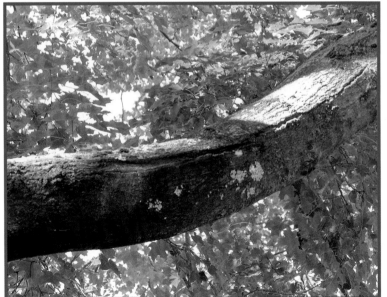

Figure 6.21 Shear plane cracks can result from bending forces on limbs or trunks that are being overloaded by gravity, snow, ice, or wind. In branches, they sometimes occur at a curve in the branch. Some shear plane cracks close and develop response growth. In this case, the likelihood of further failure of this branch within one year was rated as *improbable*.

Cracks and Decay

Sometimes, you will see cracks associated with internal decay issues. Cracks can result when thin shell walls fail in compression or torsion. You may see cracks on one side or multiple sides of a trunk or branch. Vertical **compression cracks** on one side most often occur on the compression side of the tree. You can easily simulate it on a banana by bending the fruit and watching one side of the skin rupture. Torsional cracks are often spiral and can occur on any side of the tree. Superficial cracks—those that do not penetrate deep into the wood—may not represent a serious structural issue.

Cracks on multiple sides suggest advanced columns of decay and a trunk that is buckling or twisting under load. Picture a barrel made of wood staves that is being squashed. As the barrel is squashed, the staves buckle outward, and cracks appear in between each stave. In these cases, the likelihood of failure should be considered *probable* to *imminent*.

The depth of the crack and presence of any surrounding void can sometimes be determined by probing through the crack opening. Extensive decay and recent, non-reinforced cracks often lead to a likelihood rating of *imminent*.

Horizontal (transverse) cracks on a tree trunk are not common. When it occurs, the wood fibers may be collapsing in compression. At this point, failure is more likely. It may take years before the component fails, or it may be a precursor to *imminent* failure.

Frost and Freeze–Thaw Cracks

Frost and **freeze–thaw cracks** result from exposure to sun during cold temperatures. **Frost cracks** can appear on any side of the trunk or branch, while freeze–thaw cracks are especially common on the south and west sides of the trunk (in the Northern Hemisphere) or the upper surface of the branches. Freeze–thaw cracks are common in climates where extreme cold periods are followed by rapid warming in the sun. In these cases, the core of the wood is frozen solid, but the external surface warms up and creates a freeze–thaw expansion sequence that cracks the outer wood surface. Frost and freeze–thaw cracks tend to have pronounced ribs and evidence of annual or periodic failure of the new callus wood. Shallow surface cracks are less significant than cracks that penetrate deep into the structure (potential shear plane cracks).

Figure 6.22 This crack likely started as a frost crack and later became a canker with decay. Note both the longitudinal and transverse cracks. Given a five-year time frame, the likelihood of failure of this tree was rated as *probable*.

Figure 6.23 (a) Seams are pronounced lines or zones of response growth, or lines of growth that have curved inward. They are typically associated with a wound or a crack. (b) If the seam masks a wound there may be an area of decay and possible strength loss behind it that can increase likelihood of failure.

Seams

Seams are lines formed where two edges of bark at a crack or wound meet. The likelihood of failure within three years at a seam is generally considered *improbable* to *possible* if there is solid wood behind the seam. Sometimes, however, the seam covers decay or old cracks. Depending on the extent of decay, exposure to loads, and weight distribution, failure at seams may be considered *possible* to *probable*.

Ribs

Ribs are longitudinal areas of new wood developing as a response to areas of structural weakness. They are usually an indication of a crack. Over time, the response growth may cover the weak area, leaving an abnormal bark pattern, rib, or bulge. If the rib has a sharp edge, the crack may be quite close to the surface. Ribs that contain active cracks should be evaluated similarly to other cracks. If the tree has fully closed the crack with new wood so that there is more of a blunted edge, then the crack may be buried deeper inside. Ribs covering over closed cracks should be evaluated similarly to seams. Ribs can run in a spiral manner up the trunk, and they may be discontinuous. This suggests a torsional crack system inside. Many species have a spiral grain that predisposes them to spiral ribs.

Figure 6.24 (a) Response growth resulting from a crack may create a rib. (b) If there is decay behind a rib, there can be an increased likelihood of failure.

Module 6 – Tree Inspection and Assessment

CASE STUDY

Assignment: Inventory of trees and shrubs on site, considering risk in a three-year time frame. Report any health or risk problems.

Targets and Site: Grounds of a stately home. Targets include landscape maintenance staff with rare occupancy.

Conditions: A yew (*Taxus* sp.) tree has a 12-inch (30 cm) diameter lateral limb with a shear plane crack about 2 feet (61 cm) long. Response growth has developed along both edges.

Analysis: Although the defect is well-defined, the likelihood of impact is *very low*, and the consequences of failure are *negligible*. This combination makes the level of risk *low*, no matter what the likelihood of failure is. The limb supports a large amount of active foliage that is contributing to the overall energy balance of the tree, and removal would leave a large wound. The crack has not been extending, and the tree is compensating for it.

Recommendation: Bring to the attention of the property manager; retain and monitor. Consider pruning to reduce branch weight.

Provided by Julian A. Dunster

Tree Architecture

Tree architecture, including tree growth and branching characteristics, can affect how a tree moves in the wind, how the load of the crown is distributed, and the tree's stability. Problems with other structural defects become more significant when tree architectural abnormalities are present.

Lean

Lean is the angle of the trunk measured from vertical. For most trees, lean develops as the tree grows away from neighboring trees or structures, toward light. Trees that lean may be stable for long periods of time, especially if they self-correct for the lean with response growth and redirected crown growth. However, leaning trees can be less stable than vertical trees because weight is unequally distributed over the trunk and root system. Trees with uncorrected leans and leaning trees on slopes are often less stable than straight trees or those with corrected leans. These conditions can be exacerbated if the soil is unstable and/or saturated.

Trees also may lean because of a partial failure of the lower stem or roots, or soil conditions that allow excessive root movement. Trees that increase in lean over a short time frame are of concern and should be examined promptly. Lifting of the soil on the side of the tree opposite the direction of lean (tension side) or a depression in the soil on the side of the lean (compression side) may indicate that the roots or soil are failing. Lean angle can be determined using several means, including interpretation of photographs or use of a clinometer, protractor, or digital level. The residual stability of a leaning tree may be determined by load tests. The likelihood of failure of a recently leaning tree is often *probable* to *imminent*.

Corrected Leans

Corrected leans (**sweeps**) are characterized by a leaning lower trunk and a top that is more upright. Trees may develop this form when growing at the edge of a group. Such trees are stable under normal conditions, but under additional loads they may be less stable than straight trees. A sweep may also develop when a tree leans due to partial failure and reaction wood growth has turned the crown upright. Trees with corrected leans are often considered to have a likelihood of failure within a three-year time frame from *improbable* to *possible*.

Bows

Bows are leans characterized by the top of the tree bending over more than the lower trunk, creating a curve. They typically represent a partial failure of the wood fibers, often as a result of snow or ice loading or high winds. The likelihood of failure is also related to the degree of taper; well-tapered trees are less likely to become bowed.

Most trees never recover from or respond to damage that results in a bow. Bowed trees should be examined for longitudinal cracks. The likelihood of bowed tree failure within three years may be *possible* to *probable* if it has not increased the stem diameter by compensation growth. If cracks are present, the likelihood of failure may be *probable* to *imminent* if the tree is subjected to loads from wind, snow, or freezing rain.

Trees also may develop a bowed form when growing at the edge of a group. Such trees may be stable under normal conditions but are less stable under additional loads than straight, vertical trees.

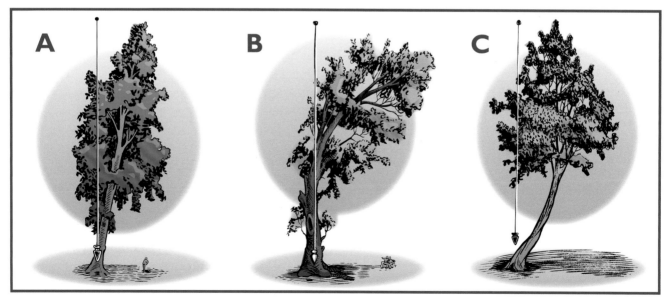

Figure 6.25 Lean is the angle of the trunk measured from vertical. The pattern of tree lean can affect the likelihood of failure. Illustrated here from left to right are (A) a straight lean, (B) a bow, and (C) a corrected lean or sweep.

One-Sided or Unbalanced Crowns

Ideally, branches are evenly distributed vertically and horizontally around the trunk. Trees with even branch distribution tend to distribute the loads evenly along the trunk, reducing the likelihood of trunk and root failure compared with trees that have uneven branch distribution. A tree may have most of its branches on one side (asymmetrical crown) if it is growing close to another tree or structure, has uneven crown dieback, is damaged by a storm, or has been pruned in that manner. In most cases, trees develop response wood to support the asymmetric crown. Asymmetric crowns can, however, contribute to failure when other problems are present, such as compromised root anchorage.

Taper, Live Crown Ratio, and Height/Diameter Ratio

Taper is the change in diameter over the length of trunks, branches, and roots, and it is important for even distribution of mechanical stress. The degree of taper is a reflection of the tree's ability to produce sufficient energy to create a uniform growth increment along the entire trunk or limb. Trees with small crowns and tall trunks are less likely to have uniform growth increments than trees of the same height but with larger crowns, because the amount of energy produced is a function of the amount of living crown.

Live crown ratio (LCR) is the ratio of the vertical extent (height) of the live crown to the height of the entire tree. Live crown ratio is affected by species, growing conditions, pruning history, previous branch failures, and natural branch shedding. A low LCR most commonly develops naturally where trees are growing in dense stands, but it can also be created by crown raising. Low LCR can be a condition of concern, especially if the tree originally developed in forest conditions and was recently exposed to higher wind conditions. A general rule is that if the LCR is less than about one-third, then there is an increased likelihood of failure, but there may be site-specific factors modifying that threshold.

Mechanical stress is also a function of tree height or branch length. A taller tree that has the same trunk diameter as a shorter tree has higher stress in the lower trunk, if all other conditions are equal, due to the longer lever arm. One way of calculating taper is to divide tree height or branch length by the trunk or branch diameter (H/D). When H/D is large (that is, the tree or branch is very slender for its height/length), the tree may be more prone to failure than when the taper has moderate or high taper with a small H/D ratio. Reported critical H/D values vary from 50:1 to 90:1, but variables affecting the actual stress include the trunk's shape, wood strength, crown configuration, and wind exposure. Therefore, H/D ratio is rarely used by itself to classify likelihood of failure.

Taper concerns do not apply to palms or to trees in an undisturbed forested area where individual trees are surrounded by other trees and are, therefore, protected from falling and the effects of wind. For some tropical tree genera with extremely hard wood (such as *Casuarina* spp.), a low LCR may not necessarily be a cause for concern.

Summary

A key part of tree risk assessment is to categorize the likelihood of failure—of one or more branches, the stem, or the roots. Visual assessment includes looking for and considering the significance of defects and structural conditions. Some structural defects or conditions are more likely to lead to failure than others. Individual defects or conditions may not by themselves indicate a serious structural problem, but when more than one defect or condition is present in a tree, their combined effect on likelihood of failure should be considered. Not all conditions and defects have a significant impact on tree structure. We must recognize that there is significant uncertainty in our assessments, and we must be realistic about the confidence level that we attach to them. The more data we have, and the greater our abilities to interpret that data, the better we can refine our predictions.

Key Concepts

1. In a tree inspection and assessment, you should consider defects, response growth, and conditions that may increase or decrease the likelihood of failure.

2. Defects include injuries, growth patterns, decay, and other conditions that reduce a tree's structural strength or stability.

3. In addition to common defects such as injuries, cracks, decay, and abnormal growth, certain branch arrangement and attachment configurations are associated with higher rates of failure.

4. The likelihood of failure associated with decay depends on the tree species, its extent and location, the presence of other defects, the expected loads, and the amount of response growth.

5. Cracks that are fresh and associated with decay, cracks occurring across the short axis of the tree part, and cracks that penetrate all the way through a branch or stem can greatly increase the likelihood of failure.

6. Because tree growth and branching characteristics can affect how loads are distributed, problems with other structural defects become more significant when tree architectural abnormalities are present.

Data Analysis and Risk Categorization

– Module 7 –

Data Analysis and Risk Categorization

Module 7

Learning Objectives

- Discuss how tree defects and conditions, site factors, and response growth affect the likelihood of tree failure.
- Explain the rating categories for likelihood for failure, likelihood for impacting a target, and for categorizing the consequences of failure.
- Describe the process of assimilating tree and site observations and categorizing the likelihood of tree failure.
- Describe the factors to consider when determining the likelihood of a failed part impacting the target and assigning one of the four likelihood of impact categories.
- Identify the factors determining consequences of failure, describe the four categories, and categorize the consequences, in a given scenario.
- Define the four risk rating categories and explain how the risk rating matrix is used to rate the level of risk based on likelihood and consequence ratings.
- Explain how clients' perception of risk and their risk tolerance affect the thresholds for action to mitigate risk.

Key Terms

acceptable risk
likelihood matrix
matrix

multiple risks
risk categorization
risk evaluation

risk rating matrix
risk perception

risk tolerance
time frame

Likelihood of Failure	Likelihood of Impact	Likelihood of Failure and Impact	Consequences of Failure	Risk Ratings
imminent	*high*	*very likely*	*severe*	*extreme*
probable	*medium*	*likely*	*significant*	*high*
possible	*low*	*somewhat likely*	*minor*	*moderate*
improbable	*very low*	*unlikely*	*negligible*	*low*

Introduction

By the time you have reached this point in the process, the time frame has been established, the tree has been examined, and you should have a good sense of whether the entire tree or a part is likely to fail. The site conditions that might affect—or have already affected—the tree are documented, and the nature of the potential consequences has been identified. The task now is to categorize the **likelihood of failure**, the **likelihood of impact**, and the severity of potential **consequences of failure** if failure and impact occurs. Once the **risk categorization** process is done, a **risk rating** can be assigned.

The Risk Categorization Process

In deriving an estimate of risk, you must consider the targets, the likelihood of a tree failure impacting a target, and the consequences of the failure. To determine the likelihood of a tree failure impacting a target, you need to consider two factors. The first is the likelihood of a tree failure occurring within a specified time frame. Examine structural conditions, defects, response growth, and anticipated loads. This is where we bring together everything we know about tree biomechanics and the implications of various defects and tree conditions that could increase or decrease the likelihood of failure, as well as certain site factors that could play a role.

The second factor is the likelihood of the failed tree or part impacting the specified target. An impact may be the tree directly striking the target, or it may be a disruption of activities due to the failure. Again, characteristics of both the tree and the site can affect the likelihood of impact.

Once you have analyzed the first two factors, you will follow a specific procedure:

1. Categorize these two factors using the **likelihood matrix** (see Matrix 1) to estimate the likelihood of the combined event: a tree failure occurring and impacting the specified target.

2. Combine the likelihood of that failure and impact with the expected consequences to determine a level of risk using the **risk rating matrix** (see Matrix 2).

3. Compare the risk rating to the level of risk that is acceptable to the client, controlling authority, or, where appropriate, societal standards.

4. Present mitigation options if the risk rating identified exceeds the level of acceptable risk.

Likelihood of Failure

Judgment about the significance of defects, site conditions, and response growth is based on the typical factors described in Modules 4, 5, and 6. Your knowledge and experience as a tree risk assessor will influence how you assess and interpret these factors. You will have to exercise judgment by using guidelines available for likelihood of failure for individual defects and conditions. You should also consider all known compounding factors as well as any response growth in the tree, which may have compensated for the conditions. Guidelines should be treated as a starting point and should be modified as needed

Figure 7.1 Risk is a combination of the likelihood of an event and the severity of the consequences. In deriving a risk rating, you must consider the targets, the likelihood of failure, the likelihood of that tree failure impacting a target, and the consequences of the failure.

so that they are appropriate for the tree and site. There may be circumstances in which you see the need to deviate or modify the guidelines or other standards that are used. Any deviations from published standards and guidelines should be noted in a report.

When more than one defect or condition is present in a tree, the impact of the individual issues as well as the combination of them must be considered. Not all conditions and defects have a significant impact on tree structure. For example, a trunk lean of 10 degrees may not be of great concern on many trees. But if there is a large, decayed root on the side opposite the lean, then the likelihood of failure increases if significant loads are likely to occur and the tree has not been compensating for the defect with adaptive growth. Assessing each condition with regard to its likelihood of failure will help you discern the significance of each condition relative to the whole tree.

Finally, consider the weather and other factors that might change the loads that the tree experiences. Tree failures usually occur when there is a critical combination of tree defect(s), conditions, and environmental forces such as wind, rain, freezing rain, or snow. Few tree failures occur in calm conditions. Most tree failures occur when wind speed exceeds the seasonal norm for the site.

Defining the Time Frame

Keep in mind that a **time frame** must be specified for the likelihood of failure rating to have meaning. The time frame is the length of time, for instance the number of years, for which the assessor is deciding whether a specific failure is likely to occur. Assessments may also be conducted using more than one time frame. For example, an assessment can be done for the next one year and for the next three years. In some cases, the results will be different. Sometimes, having assessments consider more than one time frame can be helpful for the tree owner/manager in making mitigation choices. The longer the time frame, the greater the uncertainty.

Time frames of one or three years are common. Time frames greater than five years are often not appropriate because the uncertainty over that period can be excessive. Long time frames can unnecessarily increase the rating for likelihood of failure, which could lead to unnecessary mitigation.

The time frame cannot be considered a "guarantee period" for the risk assessment.

Using the **matrix** approach as presented later in this module, the likelihood of failure can be categorized using the following guidelines:

Imminent: Failure has started or is most likely to occur in the near future, even if there is no significant wind or increased load. This is an infrequent occurrence for a risk assessor to encounter and may require immediate action to protect people from harm. The imminent category overrides the stated time frame.

Probable: Failure may be expected under normal weather conditions within the specified time frame.

Possible: Failure may be expected in extreme weather conditions, but it is unlikely during normal weather conditions within the specified time frame.

Improbable: The tree or tree part is not likely to fail during normal weather conditions and may not fail in extreme weather conditions within the specified time frame.

Tree defects and conditions are typically considered individually when assessing single trees, but sometimes they can be considered in aggregate as long as the risk being assessed is clearly defined that way. For example, the likelihood of failure of a specific dead branch might be rated as *possible*, while the likelihood of failure of any one of several dead branches in a tree might be rated as *probable*. Similarly, two or more modes of failure (codominant stems, dead branch, etc.) might be rated in aggregate, although this is more complex to consider. As always, it is essential to define the risk that is being assessed.

Figure 7.2 Imminent: The likelihood of failure of the left codominant stem is *imminent* due to the open and active crack in the union. Failure has started or is most likely to occur in the near future, even if there is no significant wind or increased load.

Figure 7.3 Probable: These conks indicate widespread and serious root decay, making the likelihood of failure *probable*. Failure may be expected under normal weather conditions within the specified time frame.

Figure 7.4 Possible: The codominant union shows significant response growth, indicating a *possible* likelihood of failure. Failure may be expected in extreme weather conditions, but it is unlikely during normal weather conditions within the specified time frame.

Figure 7.5 Improbable: This seam from frost cracking is an example of *improbable* likelihood of failure. The tree or tree part is not likely to fail during normal weather conditions and may not fail in extreme weather conditions within the specified time frame.

Likelihood of Impacting a Target

The second factor to be considered is the likelihood of the failed tree or tree part impacting the target. To estimate this likelihood, you must estimate or otherwise determine the occupancy rate of any targets within the target zone, the likely direction of fall, and any factors that could protect the target from impact by the falling tree or tree part. For example, a target may be directly under a broken branch but protected from impact by other trees or man-made structures. In this example, the occupancy rate might be constant, but the likelihood of impact might be *very low* because of the protection factors.

The likelihood of impacting a target can be categorized using the following guidelines:

- **High**: The failed tree or tree part is likely to impact the target. This is the case when there is a constant target, with no protection factors, and the direction of fall is toward the target.

- **Medium**: The failed tree or tree part could impact the target, but is not expected to do so. This is the case for people in a frequently used area when the direction of fall may or may not be toward the target. An example of a *medium* likelihood of impacting people would be passengers in a car traveling on an arterial street (frequent occupancy) next to the assessed tree with a large, dead branch over the street.

- **Low**: There is a slight chance that the failed tree or tree part will impact the target. This is the case for people in an occasionally used area with no protection factors and no predictable direction of fall; a frequently used area that is partially protected; or a constant target that is well protected from the assessed tree. Examples are vehicles on an occasionally used service road next to the assessed tree, or a frequently used street that has a large tree providing protection between vehicles on the street and the assessed tree.

- **Very low**: The chance of the failed tree or tree part impacting the specified target is remote. Likelihood of impact could be rated *very low*

Occupancy Rates

Constant Occupancy

A target is present at nearly all times, 24 hours a day, 7 days a week.

Frequent Occupancy

The target zone is occupied for a large portion of a day or week.

Occasional Occupancy

The target zone is occupied by people or targets infrequently or irregularly.

Rare Occupancy

The target zone is not commonly used by people or other mobile/movable targets.

if the target is outside the anticipated target zone or if occupancy rate is rare. Another example of a *very low* likelihood of impact is people in an occasionally used area with protection against being struck by the tree failure due to the presence of other trees or structures between the tree being assessed and the targets.

Usually targets are assessed on an individual basis, but they can also be combined to provide the client with a better perspective of what would happen in case of a failure. Consider, for example, a straight tree with considerable root decay in a courtyard surrounded on three sides by a house, a garage, and a workshop, at various distances from the tree. The house, which is only partially within the target zone, can be considered as an individual target with a *medium* likelihood of impact. Another approach is to consider all three targets in aggregate: as a "structure." In that case, the likelihood of impact would be *high*, since the structures surround the tree on three sides and, if the tree were to fail, it would probably impact one of the structures. This situation is a good example of the need to clearly define what risk is being rated.

Module 7 – Data Analysis and Risk Categorization

Figure 7.6 High: The failed tree or tree part is likely to impact the target. The likelihood of a branch failure from this tree impacting a vehicle below is *high* because the tree is near an intersection of two busy streets.

Figure 7.7 Medium: The failed tree or tree part could impact the target, but is not expected to do so. The likelihood of a branch impacting a pedestrian on this walkway is *medium* because the occupancy rate is occasional to frequent.

Figure 7.8 Low: There is a slight chance that the failed tree or tree part could impact the target. The likelihood of this tree impacting a vehicle or pedestrian is *low* because the road shows limited use and the fall direction may not be toward the road.

Figure 7.9 Very low: The likelihood of this branch impacting a specified target is remote. The site is rarely used and has very low traffic.

Matrix 1. The likelihood matrix used to estimate the likelihood of a tree failure impacting a specified target.

Likelihood of Failure	Likelihood of Impact			
	Very low	**Low**	**Medium**	**High**
Imminent	Unlikely	Somewhat likely	Likely	Very likely
Probable	Unlikely	Unlikely	Somewhat likely	Likely
Possible	Unlikely	Unlikely	Unlikely	Somewhat likely
Improbable	Unlikely	Unlikely	Unlikely	Unlikely

Categorizing Likelihood of a Tree Failure Impacting a Target

After determining the likelihood of failure and the likelihood of impacting a target, you can categorize the combined likelihood of a failure impacting a target. Use Matrix 1 for determining the **likelihood of failure and impact**. The resulting terms (*very likely, likely, somewhat likely,* and *unlikely*) are then used to categorize the tree risk rating using Matrix 2, the risk rating matrix.

An example of determining the likelihood of a failure impacting a target is as follows:

A mature tree with a large, dead branch is growing next to a one-story house. The dead branch is on the side of the tree away from the house. The likelihood of a dead branch failure within the next year was classified by a tree risk assessor as *probable*. The house is a static target with a constant occupancy rate. However, the likelihood of the branch falling from the opposite side of the tree through the rest of the tree to the house is *very low*. This results in a likelihood of failure and impact rating of *unlikely*.

Consider now that there is a parking area located partially under the branch and there are no lower branches that would mitigate the fall of the branch. A car is parked under the tree for 14 hours per day, and the driver is present for a minute or two each day as she walks between the house and the car.

Matrix 2. Risk rating matrix showing the level of risk as the combination of likelihood of a tree failing and impacting a specified target, and the severity of the associated consequences.

Likelihood of Failure & Impact	Consequences of Failure			
	Negligible	**Minor**	**Significant**	**Severe**
Very likely	Low	Moderate	High	Extreme
Likely	Low	Moderate	High	High
Somewhat likely	Low	Low	Moderate	Moderate
Unlikely	Low	Low	Low	Low

Thus, the human occupancy rate in the target zone is rare, and the car occupancy rate is frequent. This leaves two scenarios to judge:

1. The rare human occupancy rate translates to a *very low* likelihood of impacting the driver. When that is combined with a *probable* failure likelihood, the combination results in the likelihood of a failure impacting the driver of *unlikely*.

2. The car occupancy rate is frequent, and there are no structures or tree parts that would deflect or impede the fall of the branch on this side of the tree. You rate the likelihood of impact *medium*. Combining the *medium* likelihood of impact with the *probable* likelihood of failure of the branch, the likelihood of failure and impact for the car becomes *somewhat likely*.

As illustrated in this example, it is not unusual to have multiple targets with different values and occupancy rates. You should consider the most important risk targets when conducting a risk assessment, and you should be able to define each risk assessed.

Categorizing Consequences of Failure

Consequences of tree failure and impact are categorized based on the value of the target and harm that may be done to it. Using the matrix approach, you need not estimate actual values or repair costs, but you must categorize the consequences. The significance of target values—both monetary and otherwise—are sometimes subjective and relative to the client. For example, a certain target may have no apparent intrinsic or market value, but it might hold tremendous sentimental value to the client. Generally, you should assess target values from the client's perspective.

The consequences of failure and impact also depend, in part, on the tree or tree part size, fall characteristics, fall distance, and any factors that may protect the target from harm. That is, if the tree fails and impacts the target, the facts that will affect the amount of harm must be considered in the rating of the consequences. Obviously, larger parts can be more destructive than smaller ones, and the force with which a strike occurs can increase with the distance or height of fall. A tree that is very close to a two-story house, for example, might cause minor damage to the roof and eaves if it uproots and falls toward the house. But the same tree could slice the house in two if it were to fail from farther away, gaining momentum in the fall.

Protection may be provided by structures that surround the people in the target zone. Protection factors can affect the likelihood of impact rating, the consequences rating, or both. If the protection factor is not strong enough to stop the impact, then the assessor should judge if it will reduce the consequences. Buildings that are substantial provide protection to people inside them, so this factor is considered in the likelihood of impact.

Vehicles provide some protection against small-diameter falling branches. However, people in a vehicle struck by a medium-size branch may be injured but are usually not killed, so the level of consequences may be reduced by the structure of the vehicle.

Trees can also cause accidents if they fall across a motorway in front of moving traffic. Because of this, consideration should be given to traffic flow, speed, street configuration, and occupancy rates within the target zone near roads and parking lots. Keep in mind that large trees that fall into traffic can also cause significant disruption, possibly blocking major highways.

The consequences of failure can be categorized using the following guidelines:

Severe consequences are those that could involve serious personal injury or death, high-value property damage, or major disruption of important activities. Examples of *severe* consequences include:

- Injury to one or more people that may result in hospitalization or death
- Destruction of a vehicle of extremely high value
- Major damage to or destruction of a house
- Serious disruption of high-voltage distribution circuits or transmission power lines

Significant consequences are those that involve substantial personal injury, moderate- to high-value property damage, or considerable disruption of activities. Examples of *significant* consequences include:

- Injury to a person requiring medical care
- Serious damage to a vehicle
- High-monetary damage to a structure
- Disruption of distribution primary voltage power lines
- Disruption of arterial traffic that causes an extended blockage and/or rerouting of traffic

Minor consequences are those that involve minor personal injury, low- to moderate-value property damage, or small disruption of activities. Examples of *minor* consequences include:

- Minor injury to a person, typically not requiring professional medical care
- Damage to a landscape deck
- Moderate monetary damage to a structure or vehicle
- Short-term disruption of power on secondary lines, street lights, and individual services
- Temporary disruption of traffic on a secondary road

Negligible consequences are those that do not result in personal injury, involve low-value property damage, or disruptions that can be replaced or repaired. Examples of *negligible* consequences include:

- Striking a person, causing no more than a bruise or scratch
- Damage to a lawn or landscape bed
- Minor damage to a structure requiring inexpensive repair
- Disruption of power to landscape lighting
- Disruption of traffic on a neighborhood street

Continuing the example from the previous section, the consequences of a medium-sized dead branch striking a house would be *minor*, the consequences of that branch striking an unoccupied, new car could be *significant*, and the consequences of its impacting a person would be *severe*. These consequences are combined with the likelihood of failure and impact to determine risk ratings.

Rating Risk

Tree risk assessment reports should include a rating of risk. A risk rating matrix (Matrix 2) is a means of combining ratings of likelihood and consequences to determine a level or rating of risk. Although there

Steps in Developing a Tree Risk Rating

1. Identify possible targets, including those identified by the client.

2. Identify tree part(s) and/or failure mode(s) —that is, the location or manner in which failure could occur.

3. Assess and categorize likelihood of failure for each part of concern within a specified time frame.
   (*Imminent / Probable / Possible / Improbable*)

4. Assess and categorize likelihood of tree/part impacting the target.
   (*High / Medium / Low / Very low*)

5. For each failure mode/target combination, categorize the likelihood of failure and impact (Matrix 1).
   (*Very likely / Likely / Somewhat likely / Unlikely*)

6. For each failure mode/target, assess consequences.
   (*Severe / Significant / Minor / Negligible*)

7. For each failure mode/target, categorize the risk (Matrix 2).
   (*Extreme / High / Moderate / Low*)

Figure 7.10 Severe: Failure of one of these trees could involve serious personal injury or death, high-value property damage, or disruption of important activities, making the consequences rating *severe*.

Figure 7.11 Significant: Failure of a branch from the large oak (*Quercus*) pictured here could cause major damage to the house or cars. Therefore, the consequences of failure would be categorized as *significant*.

Figure 7.12 Minor: If this tree failed, it would impact the fence and smaller branches could reach the road, which could cause some vehicle damage or traffic disruption. This is an example of *minor* consequences of failure.

Figure 7.13 Negligible: There are no targets of significance within the target zone of this tree, so the consequences of failure are *negligible*. The greatest loss in this situation would be the tree itself.

are many possible approaches to risk assessment, the matrix approach was selected for use in this manual because of its broad acceptance, ease of use, and effective application for rating tree risk. This matrix was designed specifically for the evaluation of risk posed by tree failures. The limitations associated with using a matrix include the inherent subjectivity and uncertainty associated with the selection of both the likelihood and consequence factors and the lack of comparability to other types of risk assessed using other means.

Levels of Risk

In the risk rating matrix, four terms are used to define levels of risk: *extreme*, *high*, *moderate*, and *low*. These risk ratings are used to communicate the level of risk and to assist in making recommendations to the owner or risk manager for mitigation and inspection frequency. The priority for action depends on the risk rating and the **risk tolerance** of the owner or manager.

Extreme: The extreme-risk category applies in situations in which failure is *imminent*, with a *high* likelihood of impacting the target, and the consequences of the failure are *severe*. The tree risk assessor should recommend that mitigation measures be taken as soon as possible. In some cases, this may mean recommending or implementing immediate restriction of access to the target zone area to avoid injury to people.

High: High-risk situations are those for which consequences are *significant* and likelihood of failure and impact is *very likely* or *likely*, or consequences are *severe* and likelihood is *likely*. This combination of likelihood and consequences indicates that the tree risk assessor should recommend mitigation measures be taken. The decision for mitigation and timing of treatment depends on the risk tolerance of the tree owner or risk manager. In populations of trees, the priority of high-risk trees is second only to extreme-risk trees.

Moderate: Moderate-risk situations are those for which consequences are *minor* and likelihood of failure and impact is *very likely* or *likely*, or likelihood is *somewhat likely* and consequences are *significant* or *severe*. The tree risk assessor may recommend mitigation and/or retaining and monitoring. The decision for mitigation and timing of treatment depends upon the risk tolerance of the tree owner or manager. In populations of trees, moderate-risk trees represent a lower priority than high- or extreme-risk trees.

Low: The low-risk category applies when consequences are *negligible*, when likelihood of failure and impact is *unlikely*, or consequences are *minor* and likelihood is *somewhat likely*. Mitigation is generally not required. Mitigation or maintenance measures may be desired for some trees, because it is sometimes possible to reduce the risk even further at very low cost, but the priority for action is low. Tree risk assessors may recommend retaining and monitoring these trees, as well as mitigation that does not include removal of the tree. Mitigation treatments may reduce risk or future risk, but the categorized risk rating is already at the lowest level.

Continuing the earlier example from the sections on likelihood and consequences:

- For the house, the risk of a medium-sized, dead branch with a likelihood of failure and impact rating of *unlikely* and consequences rating of *minor* would result in a risk rating of *low*.

- For the parked car, the likelihood of failure and impact is *somewhat likely* and the consequences are *significant*, so the risk is *moderate*.

- For the driver of the car, the likelihood of failure and impact is *unlikely* and the consequences *severe*, so the risk is also *low*.

The overall tree risk rating would be *moderate*, the highest of these three individual ratings. Whether the clients choose to mitigate the risk depends upon their perception of risk and what level of risk they find acceptable, as well as the cost, aesthetics, and inconvenience of mitigation.

Module 7 – Data Analysis and Risk Categorization

Figure 7.14 Extreme: This tree is showing signs of root failure and it is leaning over carnival attractions. It has an *imminent* likelihood of failure, *high* likelihood of impacting the target, and *severe* consequences of failure.

Figure 7.15 High: This tree has significant decay in the lower trunk, and the likelihood of failure is *probable*. The likelihood of impacting the house is *high* because of the lean. The consequences of failure are *significant*.

Figure 7.16 Moderate: The likelihood of failure for one of the codominant stems is *probable* and the likelihood of impacting a person below is *medium* due to the moderate use of the site. The consequences of failure would be *severe* for people under the tree.

Figure 7.17 Low: Likelihood of failure is *possible*. With a likelihood of impact of *medium*, the likelihood of failure and impact is *unlikely*. Consequences of failure are *severe*.

Multiple Risks

As seen in the examples above, most trees have more than one potential failure mode and may have multiple risk targets. For example, a tree with excessive root decay may also have several dead branches; the whole tree could fail from root decay, and dead branches may fail. Similarly, the whole tree may fall on a house, while the dead branches would fall only on its driveway. When evaluating individual trees, it is appropriate to evaluate each of the factors as independent events and to recommend mitigation options along with estimated residual risks for each factor.

Risk aggregation is the consideration of **multiple risks** in combination, and it is difficult to do even with complex mathematical analyses. You *cannot* simply add or multiply the risk ratings for the individual failure modes to reach a whole-tree risk rating. As described previously, however, it is possible to group targets and/or failure modes in the definition of the risk, and to categorize the risk for the group. One example is the aggregation of many pedestrians when considering the likelihood of impacting a person on a busy sidewalk. In such a situation, you would not be considering the likelihood of impacting a specific person, but rather the likelihood of impacting *any* person. Your assessment report should clearly define what risk(s) have been rated.

Highest Risk

If you cannot add the various risks posed by a tree using the matrix methodology, how can you determine an overall risk rating for a tree? You can identify—among all the failure modes and consequences assessed—the failure mode having the greatest risk and report that as the tree risk rating. Assigning a tree risk rating for the overall tree may be useful, especially when assessing a population of trees. For example, in a given situation, whole-tree failure may be *unlikely*, but it could have *significant* consequences if it occurs; using Matrix 2, the risk rating is *low*. At the same time, failure of a dead branch may be *very likely* but have *minor* consequences; the risk rating is *moderate*. The risk rating for the tree may be reported as *moderate*, the higher of the two ratings. This rating often is presented as the single risk level for the tree, especially when dealing with populations of trees. Tree risk assessors working with a population of trees may use this overall tree risk rating to prioritize remedial action. The risk rating should be recorded with the associated source of the risk (the target/failure mode combination).

Assigning a tree's overall risk at the highest level of risk for the various target/failure mode combinations for that tree is a practical method for tree risk management. Keep in mind that the overall tree risk rating should be modified, if appropriate, following mitigation measures.

Risk Evaluation

Risk evaluation is the process of comparing the assessed risk against given risk criteria to determine the significance of the risk. This is usually done by the tree owner or risk manager, sometimes in consultation with the risk assessor. You, as the risk assessor, present the level of risk that you determined, your recommended mitigation actions or options, and their associated residual risks. (Mitigation and residual risk are presented in Module 8.) The tree owner or risk manager must then decide on what actions, if any, to take.

Risk Perception and Acceptable Risk

Tree risk assessors assess and categorize an individual tree's risk. How people perceive risk—their **risk perception**—and their need for personal safety are both inherently subjective; therefore, risk tolerance and action thresholds vary among clients. What is within the tolerance of one person may be unacceptable to another.

It is impossible to maintain trees completely free of risk—some level of risk must be accepted to experience the benefits that trees provide. **Acceptable risk** is the degree of risk that is within the client's or controlling authority's tolerance or that is below a defined threshold. Some countries have a framework directive that defines thresholds of risk tolerance and acceptability. Municipalities, utilities, and property managers may have laws, ordinances, or risk management plans that define the level of acceptable risk.

Figure 7.18 How people perceive risk and their need for personal safety is inherently subjective; therefore, risk tolerance and action thresholds vary among clients. Acceptable risk is the degree of risk that is within the client's or controlling authority's tolerance or that is below a defined threshold.

Safety may not be the only factor used by the risk manager to establish acceptable levels of risk; budget, a tree's historical or environmental significance, aesthetics, and other factors also may come into the decision-making process.

For extreme-risk trees, you should notify the owner/manager as soon as possible and, in some cases, recommend or implement immediate restriction of access to the target zone to avoid injury. For lower levels of risk, however, some discussion is usually required to understand the client's risk tolerance and determine appropriate mitigation treatments. In considering risk and mitigation measures, you should communicate the benefits of trees as well as the consequences of losing them.

A sense of perspective and context is crucial. While all trees will eventually fall down, that does not mean that all large and older trees must be cut down before they fall. There will be situations where the information available leads you to rate the risk as *high* and recommend removal if other options for mitigation cannot reduce the risk to the level acceptable to the client. There will also be situations where, after careful consideration of the information available, you rate the risk as *low* or *moderate*, and recommend retention—perhaps only after certain mitigation measures have been implemented.

What If the Situation Is Not Clear?

Sometimes, the information available will be sufficient to make a straightforward assessment of risk and develop recommendations about how to proceed. Other times, you will be confronted with ambiguity, contradictions, and significant uncertainty. How you deal with this is critical in developing a credible risk rating.

Many times, the information collected will be simple to interpret: branches or trunks with obvious, well-defined cracks; cracked trunks around a large cavity; the presence of aggressive pathogens in a well-advanced stage; heaved soil and shattered roots. These are all obvious problems, and their implications for risk are generally simple to assess. The task gets more difficult when the information is less obvious or less well-defined.

At times, you will recommend a higher level of assessment to gain more information. When considering advanced assessment techniques, it is not enough to know that there are ways to test for structural and biological conditions in a tree. You also need to know the limitations, strengths, and weaknesses of each test, approach, or way of interpreting data. And you need to be able to relate all of this to how you derive and defend the results. How much decay is too much? How many structurally sound roots are necessary for stability? How many more years can the tree or component part be retained before the risk levels become unacceptable? How much time and money should

Tree Risk Assessment Manual

CASE STUDY

Assignment: None. Public observation. Limited visual assessment.

Targets and Site: A 42-inch (107 cm) diameter pin oak (*Quercus palustris*) was located in a public area. A road curving around the tree on three sides had heavy foot and vehicular traffic. A bench was located at the tree's base.

Conditions: A quick look at the tree readily revealed that the trunk had a 10-foot (3 m) crack on the side opposite a large limb with heavy foliage that was completely exposed to the wind.

Analysis: Failure was assessed to be *imminent*, a rare occurrence except in a post-disaster situation. The likelihood of impacting a target was considered to be *medium*, making the likelihood of failure and impact *likely*. Consequences were categorized as *severe*, making this a high-risk situation.

Recommendation: "In this particular case, I sought the horticulturist (they have no arborist), explained the situation, and gave him my card to establish my credentials. I informed him that, in my best professional judgment, the tree had already started to fail and needed to be removed as soon as possible—and the bench moved that very day. I stressed that I was not trying to make a sale."

Provided by Jerry Bond

be spent chasing additional information, and, even if you collect a lot more, will it be enough to confidently categorize the risk and, perhaps, ensure acceptable risk levels?

As you gain knowledge and experience, your confidence and ability to analyze and interpret observations and data will grow.

- Be sure you understand the nature and implications of the data available. The ability to "read" the tree and interpret the data increases with both knowledge and experience, so avail yourself of information and research as it becomes available, and practice applying it in the field.

- Be sure your experience and abilities are sufficient to effectively analyze and interpret the data. There will be times when you need to bring in outside expertise.

- Always be aware of what you don't know and how that might affect your ability to analyze the information. Despite your best efforts in collecting and interpreting data, there will always be things you cannot observe and information that is unavailable at the time of the assessment. Include your observations and measurements, and document the limitations of the information available. In some cases, you may want to reserve judgment until additional information has been obtained and analyzed.

Summary

The observations you make and the data you collect in a qualitative tree risk assessment must be analyzed and assimilated into a risk rating. The steps for each identified mode of failure include evaluation of the likelihood for each part to fail, evaluation of the likelihood of the failed tree/part(s) impacting a target, categorization of the consequences of failure, and designation of a level of risk using the risk rating matrix. The risk rating is then evaluated against the client's risk tolerance.

Keep in mind that your ability to "read" the tree and interpret the data will increase with both knowledge and experience. Know the limitations of your experience and abilities to analyze and interpret the data, and know when to bring in additional expertise. Include your observations and measurements, and document the limitations of the information available.

Key Concepts

1. To determine the likelihood of a tree failure impacting a target, you need to consider two factors: the likelihood of a tree failure occurring within a specified time frame, and the likelihood of the failed tree or part impacting the specified target.

2. In discussing likelihood of failure, you should reference a time frame to put the likelihood in context. This time frame cannot be considered a guarantee period for the risk assessment.

3. Consequences of tree failure and impact are categorized based on the value of the target and harm that may be done to it. The significance of target values—both monetary and otherwise—can be subjective and relate to the client's desires and values.

4. The risk rating is determined by combining the likelihood of a failure impacting a target with the expected consequences using Matrix 2.

5. Acceptable risk is the degree of risk that is within the client's or controlling authority's tolerance, or that which is below a defined threshold.

6. For extreme-risk trees, you should notify the owner/manager as soon as possible and, in some cases, immediately restrict access to the target zone to avoid injury.

Mitigation

– Module 8 –

Mitigation

Module 8

Learning Objectives

- Discuss tree risk mitigation options, including both tree-based and target-based options.
- Prescribe mitigation options for identified risks.
- Describe how the mitigation options will reduce the identified risk issues.
- Assess the expected residual risk following mitigation actions.
- Explain the considerations in recommending options to a tree owner/manager.

Key Terms

brace rod (rigid brace)	mitigation	residual risk	timeline
cable (flexible brace)	mitigation options	retain and monitor	tree-based actions
guy	mitigation priority	structural support system	tree growth regulator
lightning protection system	prop	target-based actions	veteran tree
	pruning	target management	wildlife habitat
	pruning cycle		

Module 8 – Mitigation

Introduction

Mitigation is the action taken to reduce risk. Measures to mitigate risk can be tree-based, to reduce the likelihood of failure or the likelihood of impact; or they can be target-based, to reduce the likelihood of impact or the consequences. Either way, the measures prescribed need to be practical and effective. For each **mitigation option**, you should also estimate the risk that would remain after the work is completed—the residual risk.

Mitigation measures are typically presented as a set of options to the tree owner or risk manager, and it is ultimately the owner's responsibility to choose which option to implement.

Target Management

Movable targets within the target zone can sometimes be temporarily or permanently relocated. Consider whether mobile targets such as pedestrians or vehicular traffic can be rerouted or restricted from entering the target zone. These are often the solutions that will have the lowest impact on the tree and are often preferred if tree preservation is a primary management goal.

Sometimes, target-based mitigation options are employed as temporary measures to immediately reduce risk until other measures can be taken. Other times, the changes are permanent. **Target management** may be limited by costs and/or the willingness of owners and managers to implement the measures.

Restricting access to the target zone should be considered for extreme-risk trees and high-risk trees. In addition, relocating movable targets or reducing the occupancy rate of mobile targets may be a viable option at any level of risk, and these options should definitely be considered for medium- and high-risk trees.

Figure 8.1 Mitigation measures are typically presented as a set of options to the tree owner or risk manager, and it is ultimately the owner's responsibility to choose which option to implement. Both target-based and tree-based options should be considered.

Figure 8.2 Movable targets within the target zone can sometimes be temporarily or permanently relocated. These solutions have the lowest impact on the tree and are often preferred if tree preservation is a primary management goal.

Mitigation Treatments

Mitigation treatments are those that, when applied, reduce the risk for tree failure and damage. In most cases, you can consider at the time of assessment which action or actions could be taken to reduce the risk. Possible actions include the following:

Tree-based actions

- Prune to remove or reduce load on parts likely to fail
- Cable, brace, or prop to provide support for weak areas
- Remove the tree (and replace if appropriate)
- Modify the site to improve conditions for the tree
- Use tree growth regulators to extend the effective period of mitigation

Target-based actions

- Move the target
- Eliminate or restrict use within the target zone

Pruning

Crown cleaning (the removal of broken, dead, dying, defective, and weakly attached branches that create an unacceptable level of risk) can be accomplished in accordance with the applicable national pruning standards and ISA's *Best Management Practices: Tree Pruning*. **Pruning** to remove a primary branch hazard, thereby significantly reducing the level of residual risk, is often the preferred treatment associated with potential branch failure when a tree is not in serious decline.

Reduction, and, to a limited degree, thinning can be performed to reduce loading from gravity, wind, or precipitation. Topping is not recommended because it creates long-term problems with weak sprouts and the entry of wood decay fungi. In some cases, pruning can be combined with the installation of a tree support system to sufficiently mitigate risk and avoid tree removal. Pruning to reduce load may be an appropriate recommendation for any level of risk.

Figure 8.3 Removal of a dead or seriously defective branch will mitigate the risk of failure.

Crown raising can eliminate lower branches that are damaging structures or obstructing safe passage for pedestrian or vehicle traffic, or that obscure signs and lines of sight. This pruning type does not necessarily reduce risk associated with tree failure. Excessive crown raising can reduce taper development and limit the tree's ability to damp the effect of dynamic wind loading, thereby actually increasing risk.

Any pruning needs to be considered in the context of the overall health and vigor of the tree and whether the tree can tolerate the pruning proposed.

Consider the ability of the tree to:

- Compartmentalize and limit decay from the pruning cuts
- Recover from the loss of leaf area
- Sprout from latent buds and develop new leaf area

Tree response to pruning is determined by species, vigor, and light exposure. Species profiles should be considered.

Installing Structural Support Systems

Structural support systems can be installed to limit movement of certain tree parts. Various types of hardware are used, depending on the goals. Examples include:

- **Cables** (flexible braces), installed in the upper crown to limit the movement of weak unions or codominant stems
- **Brace rods** (rigid braces), installed close to or through weak unions, or through split sections
- **Guys**, installed to improve anchorage and stabilize lean
- **Props**, installed to support some leaning trees and low branches from below

Although structural support measures can reduce the likelihood of failure, especially of specific parts, they can also change how a tree responds to wind loading. When recommending the installation of support systems, consider not only the reinforcement of specific tree parts but also the overall effects on the tree. For instance, installing a number of cables or rods can decrease the ability of the tree crown to absorb the force of the wind and thus increase the load on the lower stem and roots. The relevant standards and ISA's *Best Management Practices* should be consulted when adopting this option.

Figure 8.4 Structural support systems can be installed to limit movement of certain tree parts. Hardware can include cables, brace rods, guys, or props.

Installing Lightning Protection Systems

In locations prone to lightning, large trees, exposed trees, trees on elevated sites, and trees close to water may be struck. A tree that is struck by lightning may suffer minimal damage, or it may be killed, structurally damaged, or even blown apart—posing a risk to any potential targets. Large and/or significant trees that are likely to be struck, or susceptible trees in locations with significant and severe consequences, can be protected from lightning damage by installing a **lightning protection system**. These systems are designed to carry the charge of a lightning strike to ground to protect a tree from significant damage, so they are more of a preventive measure than a treatment. Installation of a lightning protection system in a tree can also minimize the risk of side flash to nearby structures. The relevant national standards and ISA's *Best Management Practices* should be consulted when adopting this option.

Tree Removal

Tree removal may be an obvious choice in some situations, but in others it should be the last and least desirable option. Tree removal should be considered for high- and extreme-risk trees that are not appropriate for retention, perhaps as wildlife habitat or as **veteran trees**, and in situations in which the target cannot be moved or restricted. Trees provide many environmental, social, and economic benefits, and risk mitigation options must balance risk reduction against loss of those benefits. Rather than cutting the tree down, often there are other options available that can be deployed singly or in combination to reduce risk to an acceptable level. Avoid unnecessarily defaulting to completely eliminating risk by removing the tree.

Figure 8.5 Installation of a lightning protection system may be a good mitigation option for large and/or significant trees that are likely to be struck, or for susceptible trees in locations with *significant* or *severe* consequences of failure.

Figure 8.6 Tree removal may be an obvious choice in some situations, but in others it should be the last and least desirable option. Tree removal should be considered for high- and extreme-risk trees.

Modifying the Site to Improve Tree Health

In some situations, the site conditions around a tree can be improved to enhance tree stability. A common site-based measure is to improve soil conditions to encourage healthy, vigorous root growth and function. Increasing tree vitality through site modification can allow time for the tree's response growth to mitigate risk of failure due to various aboveground and belowground defects and conditions. Over time, development of new roots adds to tree stability, and development of new wood strengthens a stem with internal decay.

Site mitigation options to reduce tree risk may include:

- Applying mulch to improve soil conditions for root growth
- Improving drainage in wet soil to reduce tree failure in wind
- Removing fill soil placed against the trunk to reduce incidence of basal trunk disease and decay
- Creating wind buffers to reduce the force of wind on a tree
- Shoring banks or slopes to reduce erosion and slope failure below a tree
- Enlarging planter openings in paved areas to allow development of basal trunk flare and structural roots
- Removing turf around the base of a tree to enhance tree root growth
- Fertilizing to correct nutrient deficiencies
- Modifying irrigation systems to avoid wetting the tree trunk during operation for species prone to root and root collar diseases
- Reducing soil compaction to allow greater root development

Site modification can be combined with site-restriction measures to reduce target occupancy. In situations of extreme risk, however, site improvement should not be the sole mitigation measure because it is unlikely that site improvement will decrease tree risk quickly. Some treatments to improve site conditions and reduce tree failure may take years or decades to be effective.

Figure 8.7 In some situations, the site conditions around a tree can be improved to enhance tree stability. A common site-based measure is to improve soil conditions to encourage healthy, vigorous root growth and function. Increasing tree vitality through site modification can allow time for the tree's response growth to mitigate risk of failure due to various aboveground and belowground defects and conditions. These measures may reduce risk in the long term, and are sometimes recommended in conjunction with other mitigation strategies.

Using Tree Growth Regulators

Tree growth regulators (TGRs) alter growth hormones within a tree and are applied to reduce the elongation of stems. When tree growth regulators are used in conjunction with pruning, it may be possible to extend the reduction in wind resistance or clearance from utilities and structures for several years. In some cases, application of certain TGRs may also increase root development, which can eventually improve tree health and increase tree stability.

Because there is a time lag between treatment and effect with TGRs, their use may be most appropriate on low- and moderate-risk trees or in combination with other mitigation measures. Use of TGRs should not be considered as an appropriate measure to achieve immediate risk reduction.

Table 8.1 Tree-based mitigation options for specific tree defects, where there are targets of concern. Also consider target-based mitigation options such as moving the target or restricting access within the target zone.

Type of Defect	Mitigation Options (may be used in combination)	Conditions and Reason
Trunk or root decay	Pruning	With low to moderate decay, prune to reduce the load and size of the crown, to reduce the stress on the weak area. Significant retrenchment (crown reduction) may be an option.
	Removal	Where decay is extensive and response growth lacking, removal may be reasonable.
	Support system	Cabling typically will not mitigate this defect. Propping may be feasible if the tree is leaning strongly.
Branch crack or decay	Pruning	For low to moderate decay, prune to reduce the load and size of the branch.
	Support system	Where damage is extensive, cable into solid wood. If the branch is cracked, installation of brace rods may be an option, in combination with cable installation to reduce movement. Cabling in decayed wood is not advised.
	Removal	If pruning and/or installation of support systems is not an option, branch removal should be recommended.
Horizontal branch with poor taper and excessive end weight	Pruning	The first choice is to prune to reduce the weight and extension of the branch.
	Support system	If pruning is inadequate to reduce stress on the branch, propping or cabling could be considered, possibly in combination with pruning. If practical, move the target and let low branches settle to the ground.

Type of Defect	Mitigation Options (may be used in combination)	Conditions and Reason
Poorly tapered trunk, recently exposed, with high height: diameter ratio and low live crown ratio	Removal	Remove the tree or move the target. Thinning will seldom mitigate this condition.
	Restrict access	Over time, vigorous trees will develop taper and branching suitable for edge conditions, and risk will decrease.
Leaning tree	Retain and monitor	Where top has grown vertically (self-correcting), no treatment may be needed.
	Pruning	For moderate leans, prune to reduce load, branch length, or tree height.
	Removal	For recent leans with mounded or cracked soil behind the lean, removal usually is warranted.
	Support system	Propping or guying could be coupled with pruning.
Weak branch attachment due to included bark, V-shaped junction, or codominant stems	Pruning	Prune to reduce weight on weakly attached branches, subordinate one of the stems, or remove one of the stems.
	Support system	Consider installing a cable/brace or propping system.
	Removal	Consider removal if pruning and/or support system installation will not sufficiently reduce risk.
Weakly attached regrowth from previous loss of top caused by storm or pruning	Pruning	Prune to reduce load and length of regrowth. Subordinate and thin regrowth to provide spacing, retaining shoots with the best attachments.
	TGR	Application of a TGR can reduce stem elongation.
Roots severed near trunk	Pruning	Prune to reduce load and size of crown (crown reduction).
	Removal	If root loss is significant and tree is exposed to strong winds, consider removal.
	Support system	Guying may be possible in some circumstances.
Dead branches and/or hangers	Pruning	Prune to clean the crown.

Creation of Wildlife Trees or Retention of Very Old Trees

In some low-use locations, dead and decaying trees may be retained for **wildlife habitat**. Selection and retention of suitable trees for wildlife habitat must take into consideration the proximity of targets in the short and longer term because it will affect risk levels. Trees for wildlife habitat with significant targets should be monitored. Ideally, such trees would be in areas that have no significant targets within the target zone or that have only targets with a rare or occasional occupancy rate.

Very old and mature trees in natural settings may reconfigure as they age and deteriorate, a process sometimes called natural retrenchment. The trunk diameter may continue to grow while branches die and fail—reducing overall height of the tree and increasing stability. Where tree risk is a concern, arborists can imitate this process by employing specific crown-reduction techniques used in conservation arboriculture. One management strategy for these trees is to maintain them at a height shorter than the distance to the nearest target. Some retrenched trees have remained stable for decades. Like wildlife trees, however, they should be monitored, and should be considered for removal or mitigation if the risk exceeds allowable thresholds.

Residual Risk

Residual risk is the risk remaining after mitigation. There is always some residual risk. With tree removal, for example, that residual risk is brought to near zero; however, even stumps can pose some residual risk, such as being tripped over. The level of residual risk should be part of the report so that an informed decision can be made regarding the mitigation options. While the highest risk factor is often the focus of remedial action, you should recommend mitigating options for each condition evaluated and estimate the residual risk following those actions. If the highest risk is mitigated, the residual risk drops to the next highest risk rated.

Residual risk is calculated the same way that the original risk is calculated except that the likelihood of failure, likelihood of impact, and consequence factors

Figure 8.8 In some low-use locations, dead and decaying trees may be retained for wildlife habitat. Selection and retention of trees suitable for wildlife habitat must take into consideration the proximity of targets in the short and long term because it will affect risk levels.

Mitigation

Guidelines for tree risk mitigation:

- Extreme-risk situations should be mitigated as soon as possible. Immediate action may be required to restrict access to the target zone.
- High-risk situations should be mitigated as soon as it is practical.
- Moderate-risk situations may not require mitigation but, if deemed necessary, could be mitigated when budget, work schedule, or **pruning cycle** allows. If the risk is acceptable to the client, the tree(s) could be **retained and monitored**.
- Low-risk trees should be retained and monitored (if appropriate) and/or mitigated, if desired, when the budget, work schedule, or pruning cycle allows.

Exercise judgment in determining **mitigation priority** within groups of trees rated at the same level of risk. Factors to be considered include: location, targets, tree/part size, and the logistics of mitigation.

are estimated for the conditions that would exist after the prescribed mitigation is completed. Often, only one of these three factors will change, so calculations are simplified. Usually, the likelihood of failure is reduced with tree-based mitigation, and the likelihood of impact is reduced with target-based mitigation.

The level of residual risk needs to be acceptable to the risk manager/owner. If the residual risk exceeds the acceptable risk level, that treatment option is not the best course of action in that specific context. Some countries follow the principle of "as low as reasonably practicable" (ALARP). To meet this principle, one must demonstrate that the cost involved in further reduction of risk would be grossly disproportionate to the benefit gained. ALARP is sometimes applied in situations where large populations of trees are being managed.

Recommending Mitigation Options

In most cases, as a tree risk assessor, you will:

- Provide a recommendation or recommendation options for reducing risk
- Provide information on residual risks associated with the mitigation
- Propose a timeline for mitigation action(s)
- Suggest priorities for the mitigation work if multiple trees are being assessed

Mitigation may involve more than one action. The tree risk rating based on the highest risk factor assigns an overall tree risk, but there are typically other factors that also should be considered and that may require additional actions if the tree is to be retained. Once the highest risk factor has been mitigated, the tree risk rating drops to the next highest risk factor. Be careful in your choice of words. It is generally unrealistic and ill-advised to suggest that a tree can be made "safe," because there is almost always some residual risk. In most circumstances, it is best to discuss mitigation of risk, not elimination.

When suggesting work such as cabling, bracing, pruning, and installation of lightning protection systems, you should also recommend that these measures be implemented in accordance with industry accepted standards, such as those of the American National Standards Institute (ANSI) or ISA's *Best Management Practices*.

Safety

With some extreme-risk trees, mitigation may involve the immediate restriction of use within the target zone before removing the tree. Safe tree removal may require using a crane or an aerial lift because additional loads on the tree from the weight of a climber or from lowering branches may cause the tree to fail.

Trees that pose a high or extreme risk to the public may pose an even higher risk for workers assigned to mitigate that risk. Before beginning work, tree

CASE STUDY

Assignment: Risk assessment of urban park trees.

Targets and Site: Urban park; row of trees along a waterfront trail. The occupancy rate is occasional because it gets moderate use during daytime only, and mainly during good weather. Because benches are available, some people may be in the target zone for extended periods.

Conditions: One tree in the row has a large, dead limb in its crown, over the trail. Close examination with binoculars reveals a cavity with an active fungal conk at the base of the dead limb. Due to its configuration, however, the limb is unlikely to strike the nearest bench.

Analysis: Likelihood of failure is *probable*, and the likelihood of impacting a person is *low*, so the likelihood of failure and impact is *unlikely*. The consequences of failure are *severe* because of the human targets. The risk rating is *low*.

Recommendation: Remove the dead limb to reduce the risk. Even though the risk is low, mitigation is easy and inexpensive. If risk tolerance is low, consider restricting access to this portion of the trail until mitigation can take place.

Provided by Julian A. Dunster

workers should assess the risk that a tree poses. It also may be appropriate, in some cases, for risk assessment reports to communicate specific hazards for tree workers who may have to work on the trees to mitigate risk.

Recommending Mitigation Priorities

With a single, privately owned tree, there is little need for prioritization of work. The owner decides what action to take and schedules the work, if any. With a population of trees, such as in a municipal or utility context, scheduling becomes more important and requires prioritization. Assessment reports often include a list of mitigation measures that encompass many trees with similar risks that would receive similar mitigation treatments. It may be possible, based on your knowledge of the many sites and circumstances, to offer additional guidance to the risk manager by suggesting priority areas and tasks.

How these priorities are developed will depend on the targets, the nature of the risks, time issues (what is most pressing?), and options for mitigation, such as short- or long-term restriction of access. Prioritization should consider any applicable ordinances or regulations. The rationale for your suggestions should be described.

An additional consideration should be the short- and long-term implications of the actions prescribed and how these affect priority. For example, whole tree removal creates a potential space for replanting; if this is done right after removal, the new tree can become established sooner. Installation of a fence to restrict access for a few months can buy some time while other issues are dealt with.

If a tree with an extreme risk is discovered, notify the owner/manager as soon as possible to take action. You will not only be taking reasonable and prudent steps to reduce the risk of injury, but you will also avoid liability in the unfortunate event of the tree failing

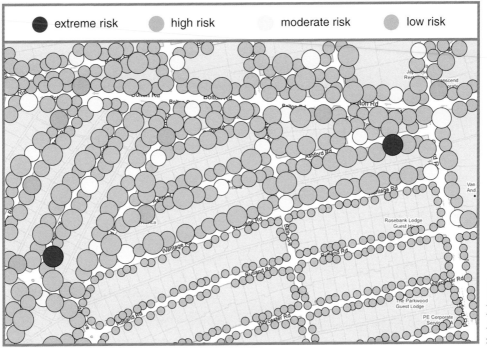

Figure 8.9 With a population of trees, such as in a municipal or utility context, scheduling is very important and requires prioritization. Assessment reports often include a list of mitigation measures that encompass many trees with similar risks that would receive similar mitigation treatments.

Tree Risk Assessment Manual

Figure 8.10 Restricting access to the target zone should be considered for extreme-risk trees and high-risk trees.

before mitigation could be implemented. In the rare event that failure is in progress, it may be necessary to notify the client and restrict access to the target zone.

There may be times you cannot provide all of the answers sought by the risk manager without spending more time and money investigating specific issues. In these situations, you can note the limitations of your report or recommend an advanced assessment, and let the risk manager decide how to proceed.

Timelines for Mitigation

The **timeline** for implementing mitigation treatments to reduce risk varies with the degree of risk present: the greater the risk, the shorter the recommended timeline.

Guidelines for tree risk mitigation recommendations:

- Extreme-risk situations should be mitigated as soon as possible. Immediate action may be required to restrict access to the target zone.
- High-risk situations should be mitigated as soon as it is practical.

- Moderate-risk situations may not require mitigation but, if deemed necessary, could be mitigated when budget, work schedule, or pruning cycle allows. If the risk is acceptable to the client, the tree(s) could be retained and monitored.

- Low-risk trees should be retained and monitored (if appropriate) and/or mitigated, if desired, when the budget, work schedule, or pruning cycle allows.

Summary

Mitigation treatments are those that, when applied, reduce the risk for tree failure and damage. In most cases, you should identify at the time of assessment what action or actions could be taken to reduce the risk. You should also estimate the residual risk for each mitigation option. With some extreme-risk trees, mitigation may involve the immediate restriction of use within the target zone before removing the tree. The tree owner/manager decides what actions to take based on their tolerance.

Key Concepts

1. Measures to mitigate risk can be tree-based, to reduce the likelihood of failure or the likelihood of impact; or they can be target-based, to reduce the likelihood or the consequences of impact.

2. Tree-based mitigation options include pruning, installation of structural support systems, lightning protection, site modification to improve growing conditions, and use of growth regulators to extend mitigation periods.

3. At times, tree removal may be the appropriate mitigation measure, but, as a risk assessor, you should avoid a default position of removal.

4. Target-based mitigation measures include moving the target or restricting access to the site. It may be appropriate in some cases to create wildlife trees or manage very old trees as veteran trees or heritage trees. These options are often combined with site restrictions when the trees are located in public areas.

5. Residual risk is the risk remaining after mitigation. The level of residual risk needs to be acceptable to the risk manager/owner. If the residual risk exceeds the acceptable risk level, that treatment option is not the best course of action in that specific context.

6. With populations of trees, mitigation may require prioritization. It may be possible, based on your knowledge of the many sites and circumstances, to offer additional guidance to the risk manager by suggesting priority areas and tasks.

7. The timeline for implementing mitigation treatments to reduce risk varies with the degree of risk present: the greater the risk, the shorter the timeline. For extreme-risk trees, you should recommend action be taken as soon as possible (that day or within the next few days). For high-risk trees, you should recommend mitigation as soon as practical, when the work schedule or pruning cycle allows.

Reporting
– Module 9 –

Reporting

Module 9

Learning Objectives

- Describe the options for effectively communicating risk issues to the client/tree owner.
- Define each risk that is being assessed.
- Explain the key components of a written report.
- Describe effective uses of verbal (oral) reports and their limitations.
- Discuss timelines for mitigation and intervals for reassessment.

Key Terms

analysis	inspection interval	reporting	work order
client	limitations	verbal report	written report
conclusions	recommendations		

Introduction

After you have conducted a tree risk assessment, the information, **conclusions**, and recommendations need to be communicated to the **client** (tree owner/manager, designated person, or agency). The preferred method is to prepare a clear and concise written report because it documents your findings.

In some instances, the report may be verbal (oral) or videographic. A combination of verbal—often at the site—and written communications can help you learn the client's response to your evaluation and recommendation and provide both you and your client with a record of what you observed and recommend. A work proposal or **work order** with a recommendation for risk mitigation also may be considered a type of risk assessment report. You should retain copies of all risk assessment reports and supporting evidence such as photographs.

In all cases, the communication process needs to be clear, unambiguous, and readily understandable to the person receiving the information. Remember that for any given tree, you may have rated the risks associated with multiple targets and conditions of concern, so it is important to define each risk that you have rated. At the end of the communication, regardless of its form, all of the key points (levels of risk, recommendations for mitigation, and the scheduling of work and reinspection) should be clearly presented to the client, tree owner, and tree workers.

When communicating with clients, keep in mind that they are not arborists and are likely unfamiliar with arboricultural or risk terms and procedures. Explain important concepts to help them understand the information you are conveying. Avoid overly complex descriptions and technical jargon that confuses rather than clarifies your report.

Written Reports

Written reports range from a simple memo or e-mail, to a work order, to a detailed report. The former will be of most use as a quick confirmation of verbal agreements: "As we discussed on site yesterday, the branch on the maple in your front yard is likely to fail and strike your house within a one-year time frame, resulting in high risk. It should be removed as soon as practical" or "Based on what we saw on site this morning, the codominant stem on the oak in your backyard is not at a high risk of failure and does not need to be removed." Even simple memos or e-mails, if used, should contain the date of the discussion, the people involved, the site location, and enough description to be clear.

Detailed written reports are the most common approach for **reporting** on a tree risk assessment. They should document a systematic thought process, **analysis**, and **recommendations** for action. They will typically contain the following:

- Name of the tree risk assessor, company name, and the date of assessment
- Statement of the assignment scope of work
- Location and/or identification of the tree(s) assessed
- Identification of the level of assessment (limited visual, basic, or advanced) and details of the method (for example, "Basic assessment, using a mallet and a probe")
- Description of targets, occupancy rates, likelihood of impacting the target, and potential consequences of failure
- Description of relevant site factors that were considered (history of failure, storm patterns)
- Documentation of the likelihood of failure, such as a list of tree conditions, structural defects, potential load, and response growth that were observed. Measurements of defects (size and shape) may be included
- Discussion of risk assessment and rating, including information on specific failure mode/target combinations, and conclusion
- Discussion of options and/or recommendations for mitigation (for example, "Move target, prune, or remove the tree"), including recommendations for timing of mitigation
- Description of residual risk
- Recommendations for reassessment (interval, level, and type)
- Identification of limitations of the assessment

For limited visual assessment of a population of trees, a complete written report should include the location of trees that had one or more of the specified defects or conditions and the type of remediation required. In addition, the report may provide work priorities, tree species, tree size, and the conditions identified. For a larger population of trees, all relevant information can be included in a multi-year, large-area management plan, if that is required by the scope of work. The management plan should also include suggested priorities for mitigation. Some reports may also include prescriptions for the creation of hazard and safe work zones, if required by local authorities.

Verbal Reports

Direct verbal communication with the client about tree risk issues is important and can be effective and timely. You should talk with clients about their risk tolerance and help them formulate a threshold at which they are comfortable. You could also discuss management options, the pros and cons of each, and ask about client preferences. Then, in the follow-up written report, briefly describe the options presented and identify and explain more fully the client's preferred option. A work order describing the risk mitigation agreed to verbally can serve as documentation.

Verbal reports allow for immediate action and feedback, eliminating the time-consuming process of writing a report and waiting for the recipient to read and consider it. If you can observe the action taken following a verbal report ("the branch has now been removed"), then the message has been received and acted on. Those actions may themselves then become part of a later written report that describes actions undertaken on site.

A verbal report can also be an important first step in informing the client about an extreme risk that requires immediate attention. In that situation, follow up the verbal report with a written report as soon as possible. That serves to (1) confirm that you have verbally informed the client about the risk issues, (2) remind him or her about these issues, and (3) clearly show that you have exercised due diligence in ensuring that the client is fully informed about the problems.

Sometimes, arborists are asked to present reports at public meetings or hearings. An example is the presentation of risk assessment results and recommendations to a municipal tree board or council. You could also be called to give testimony in court or through a legal deposition, based on an assessment job you have done, or as an expert witness on a liability case. These possibilities underscore the need for thorough documentation of your work. Moreover, your skills in communicating of your process, observation, analysis, and recommendations might very well make the difference in how your work is perceived.

Verbal reports are quick and simple and can be effective, but they have limitations. Later on, one or both parties may disagree about what was said or may feel that the emphasis on key points was not clear enough—hence, that certain important actions were missed. Preparing a follow-up e-mail or memo that provides a brief written summary of what was discussed and decided can minimize this limitation.

Figure 9.1 You should talk with clients about their risk tolerance and help them formulate a threshold at which they are comfortable. You could also discuss management options, the pros and cons of each, and ask about client preferences.

Figure 9.2 You can use software to crop, adjust exposure, and annotate images. Under no circumstances, however, should the images be altered to change their content.

Video or Photographs

The use of video or photographs to document risk issues can be a useful form of additional information and data, especially when forensic investigations into tree failure and residual risk issues arise. Good application of these techniques requires an understanding of equipment, photographic quality, and storage. Your goal with video or photographs is to provide clear, well-exposed, and sharply focused images that illustrate the point you wish to make.

The technology to make videos and photographs is relatively inexpensive, widely available, and simple to use. However, obtaining good-quality images takes some skill and understanding of the equipment, and you still need to be systematic, thorough, and able to clearly portray the issues of interest through this medium. If you intend to use this approach, be sure to practice with the equipment beforehand, understand light conditions, and be aware of the limitations of the equipment.

On site, the use of objects of known size, such as a ruler, lens cap, or a measured length of flagging tape, can be a helpful way to demonstrate the scale of important conditions. Placement of flagging tape at known heights aboveground helps to establish vertical scale or to provide a location in a landscape.

Take notes to supplement the video or photographs, and be sure to note which issues apply to which tree.

Back in the office, label photos and videos systematically. Using software to crop, correctly expose, and annotate images can be very helpful. Under no circumstances, however, should the images be altered to change their content. Aspects such as contrast, exposure, and color balance can be adjusted if that helps to accurately depict the scene documented. In some situations, you may be required to hand over all original images as soon as they are taken to establish a chain of evidence. You would then get copies back for use in your reports. Some camera manufacturers offer image authentication software that locks in data about the image and subsequent adjustments.

Limitations of Tree Risk Assessment

Limitations of tree risk assessments arise from uncertainties related to trees, defects, and the loads to which they are subjected. You should include the limitations of your assessments in your risk assessment report, including the limitations of the methodology used and any limitations related to the ability to access or assess the tree, site, or potential targets.

Some of the limitations that are common to risk assessment reports include the following:

- Tree risk assessment is limited in scope to the specific risk(s) of interest, and does not include any and all risks.
- Tree risk assessment considers significant known and/or assigned targets and visible or detectable tree conditions.
- Tree risk assessments represent the condition of the tree and site at the time of inspection.
- Not all defects are detectable and not all failures are predictable.

- The time frame for risk categorization should not be considered a "guarantee period" for the risk assessment.
- Only those trees specified in the scope of work were assessed, and assessments were performed within the limitations specified.
- Any tree, whether it has visible weaknesses or not, will fail if the forces applied exceed the strength of the tree or its parts.

Inspection Intervals

Do not confuse the inspection interval with the time frame established for likelihood of failure. **Inspection interval** is the time between assessments. Site and tree conditions change over time, so when reinspection is justified by the level of risk or target value, you should include a recommendation that risk assessment recur on a regular basis.

Timing of the initial risk assessment and frequency of future assessments are often not at your discretion. However, after a tree has undergone the initial assessment, you should recommend an inspection frequency based on the level of risk and the goals of the client. The inspection interval typically ranges between one and five years, but it may be more or less often, depending on the age of the tree, level of risk or residual risk, specific conditions, client goals and resources, or regulations. For example, a low-risk tree may be assigned an inspection interval of three to five years, or perhaps more. For a high-risk tree, an inspection period of several months to one year may be recommended. Generally, it is a good idea to inspect trees with known structural weaknesses and/or high-value targets after major storms or other exceptional events on the tree site (like forest clearing, trenching, or other construction work) to identify damage or changes in condition that may have occurred.

The time between tree risk assessments conducted by or on behalf of municipalities, utilities, and other entities that manage large populations of trees is often defined by risk managers acting on behalf of those agencies or their controlling authority. Public agencies, utilities, and large-property managers may identify zones of similar tree population, site use, and facility type. The inspection interval for each zone is then specified. Zones of higher priority should be inspected more frequently than zones with lower priority.

Scheduling assessments in a specific season can aid the assessment. For instance, the crowns of deciduous trees are more easily assessed for failure likelihood in the winter than in the summer when in full leaf because the branch structure and unions are more easily seen. If you have identified a particular decay fungus that produces annual fruiting bodies (such as conks, brackets, mushrooms) at specific times of year but the structures subsequently degrade or disappear, evaluating the progress of decay will be easier if the inspection is scheduled to correspond with the emergence of the fruiting bodies. Staggered inspection intervals, such as every 8 or 16 months, will allow you to see the trees in different seasons. For the utility arborist, assessments could be combined with periodic line clearance pruning operations or conducted as separate activities. Commercial arborists may include assessments as a part of a Plant Health Care program. Municipal arborists may include inspections as a component of cyclic pruning programs or in response to calls from citizens about a particular tree or group of trees.

Figure 9.3 Communicating the results of the risk assessment to the client is a critical component in the risk assessment process that can involve arboriculture, public safety, and potential legal issues. There are several ways to present this information, but the preferred approach is some form of written communication. A detailed formal report is a sign of professionalism.

Module 9 – Reporting

CASE STUDY

Assignment: Inspect a large pin oak (*Quercus palustris*) that recently failed and caused personal injury to a driver in a passing vehicle. Determine the cause of failure, and report whether the failure should have been foreseen when the tree had been pruned six months prior.

Targets and Site: Near the intersection of two lightly-used roads. Targets included passing vehicles, pedestrians, parked cars, and the front porch of an adjacent home, all with occasional occupancy.

Conditions: The tree had failed the previous day, during an approaching storm. Winds were gusting to 60 mph. It failed at the base due to extensive root decay that had progressed up into the root collar area. There were no fruiting bodies, cavities, or external signs of decay. Response growth was minimal and absent in several places around the trunk. The tree showed good vigor for its age and species.

Analysis: The extensive basal and root decay would not have been visible during a routine visual inspection. It would have been detected with advanced assessment techniques, but there was no indication of the need for advanced assessment other than the general size and maturity of the tree. Moreover, the tree failed in very strong wind conditions.

Report: The report for this case needed to be extremely thorough, including strong documentation. The case was very likely to go to court with high stakes. Many measurements and photographs were taken and included in the report. The report was very clear in describing the scope of the assessment, the tree and site conditions observed, and the limitations of assessing risk factors after failure.

Note: *The case did eventually go to court nearly two years later, and the clear documentation of the details in the written report were essential to the testimony.*

Provided by E. Thomas Smiley

Finally, bear in mind that all assessments represent probabilities that become more uncertain over time because of changes in conditions. In other words, the longer the inspection interval, the less reliable the assessment will be. It is therefore important to communicate to the client the time-dependency of the assessment, something most easily done when recommending the inspection interval.

Summary

Communicating the risk assessment results to the client is a critical step in tree risk assessment. The results and recommendations form the basis for subsequent risk management activity. There are several ways to accomplish this, but the preferred approach is some form of written communication. Whatever method is selected, the report must be clear and unambiguous.

Key Concepts

1. After you have conducted a tree risk assessment, the information, conclusions, and recommendations need to be communicated to the client. The preferred method is in a clear and concise written report because it documents your findings and recommendations.

2. Some of the key elements in a risk report are the location or identification of the tree(s) assessed, level of assessment, targets, risk rating, options and/or recommendations for mitigation, residual risk, and limitations of the assessment.

3. Direct verbal communication with the client about tree risk issues is important and can be effective and timely. You should talk with clients about their risk tolerance and help them formulate a threshold at which they are comfortable.

4. Verbal reports are quick and simple and can be effective, but they have limitations. Later on, one or both parties may disagree about what was said or may feel that the emphasis on key points was not clear enough—and that certain important actions were missed.

5. Limitations of tree risk assessments arise from uncertainties related to trees, defects, and the loads to which they are subjected. You should include the limitations of your assessments in your risk assessment report, including the limitations of the methodology used and any limitations related to the ability to access or assess the tree, site, or potential targets.

Appendix 1
Using the ISA Basic Tree Risk Assessment Form

This form is provided with the *ISA Tree Risk Assessment Manual* and is intended to act as a guide for collecting and recording tree risk assessment information. This form is for trees receiving a basic (Level 2) risk assessment. It is *not intended for use* with limited visual (Level 1) or advanced (Level 3) assessments. Space is provided to write comments and notes for various conditions that are not included elsewhere on the form or for points that need additional explanation. *It is not necessary to mark every box or to fill in every line on this form.* Only information relevant to the tree risk assessment should be collected. You may adapt this form for your specific needs or you may use your own method of collecting and analyzing field data.

PAGE 1—DATA COLLECTION

Section 1—Assignment and Tree ID

Client _____ Date _____ Time _____
Address/Tree location _____ Tree no. _____ Sheet _____ of _____
Tree species _____ dbh _____ Height _____ Crown spread dia. _____
Assessor(s) _____ Tools used _____ **Time frame** _____

This section outlines the basic information for your assessment. This will be valuable information when drafting your written report. Be sure to refer back to the time frame stated in this section when determining likelihood of failure later on this form.

Client—name of the person who hired you to perform the assessment or agency for which you are working.

Date—date of the tree inspection.

Time—time of the tree inspection.

Address/Tree location—the physical address, GPS coordinates, or other location description of the tree and the location of the tree on the property, such as "backyard" or "between street and sidewalk on the north side of walk." A typical entry may be "411 Pine Street, Oakville. Large tree on left near driveway."

Tree no.—if the tree has an inventory tag with a number, it should be entered here. If a group of trees without tags are assessed, they may be assigned a sequence number.

Sheet—if multiple sheets are used for a tree assessment—or if a group of trees are assessed—the sheet number and total number of sheets used on the job may be entered.

Tree species—include the common and/or scientific name of the tree; cultivar, if known.

dbh—diameter at breast height [U.S., 4.5 feet (1.37 m); or customary diameter measure for your country; IUFRO standard is 1.3 m above ground] measured in inches or centimeters.

Height—tree height either visually estimated or measured. If measured, the tool used for this measurement should be noted in Tools used.

Crown spread dia.—average diameter of the drip line of the tree; measured or estimated.

Assessor(s)—name of the person or people collecting the tree risk information; may also include qualifications such as "TRAQ."

Tools used—list of tools used in the assessment such as "mallet" or "binoculars." If no tools were used, write "none" or leave blank.

Time frame—period in which you are estimating the likelihood of failure, typically between one and five years. Time frame is essential when rating the likelihood of failure with all categories except *imminent*, which has a different time frame (very soon).

Section 2—Target Assessment

Target Assessment								
Target number	Target description	Target protection	Target zone			Occupancy rate 1 – rare 2 – occasional 3 – frequent 4 – constant	Practical to move target?	Restriction practical?
			Target within drip line	Target within 1 x Ht.	Target within 1.5 x Ht.			
1								
2								
3								
4								

The Target Assessment chart is used to list target(s)—people, property, or activities that could be injured, damaged, or disrupted by a tree failure—within the striking distance (target zone) of the tree part concerned. Four lines are provided; additional targets can be listed on a separate form. Target information will correspond with the Risk Categorization chart on the back of the form.

Target number—many trees have multiple targets within the target zone; the target number is provided to list individual targets and to facilitate inclusion of this number in the Risk Categorization chart so that the target description does not need to be rewritten.

Target description—brief description such as "people near tree," "house," "play area," or "high-traffic street." Location of the target can be noted by checking one of the distance boxes under Target zone.

Target protection—note any significant factors that could protect the target because this may affect the likelihood of impact and/or the consequences of failure.

Target zone—identify where the targets are in relation to the tree or tree part:

Within drip line—target is underneath the canopy of the tree.

Within 1 × Ht—target is within striking distance if the trunk or root system of the tree fails (1 times the height of the tree).

Within 1.5 × Ht—target is within striking distance if the trunk or root system of the tree fails and there are dead or brittle branches that could shatter and fly from the failed tree.

Occupancy rate—an estimated amount of time the target is within the target zone. Use corresponding numbered codes (1–4):

1. **Rare**—the target zone is not commonly used by people or other mobile/movable targets.

2. **Occasional**—the target zone is occupied by people or other targets infrequently or irregularly.

3. **Frequent**—the target zone is occupied for a large portion of the day or week.

4. **Constant**—a target is present at nearly all times, 24 hours a day, 7 days a week.

Practical to move target?—check box if it is practical to move the target out of the target zone if mitigation is required.

Restriction practical?—check box if it is practical to restrict access to the target zone.

Section 3—Site Factors

Site Factors

History of failures _____ **Topography** Flat☐ Slope☐ _____% **Aspect** _____

Site changes None☐ Grade change☐ Site clearing☐ Changed soil hydrology☐ Root cuts☐ Describe_____

Soil conditions Limited volume☐ Saturated☐ Shallow☐ Compacted☐ Pavement over roots☐ _____% Describe _____

Prevailing wind direction _____ **Common weather** Strong winds☐ Ice☐ Snow☐ Heavy rain☐ Describe _____

Site factors may influence the likelihood of tree failure. This section provides a list of common site factors that should be considered. There may be other site factors that are critical on a given site or that you should note even if they are not on this form. Any of these factors can be further described in the space provided or on additional paper. Other site factors affecting wind load should be noted. These may include the site elevation, surface roughness, and hilltop locations.

History of failures—note and describe evidence of previous whole-tree failures on the site, and estimate the time frame for how recently they occurred. Previous branch failures should be noted in the Crown and Branches box (located in the Tree Defects and Conditions Affecting the Likelihood of Failure section of the form).

Topography—check boxes for flat or sloping topography; an estimate of the slope percentage may be included.

Aspect—the compass direction that the slope is facing.

Site changes—factors affecting the root system of the tree or the change in exposure of the tree to wind. Check all that apply:

None—no evidence of recent site changes.

Grade change—soil was added or removed from the site.

Site clearing—adjacent trees, which may have blocked the wind, have been removed or significantly reduced.

Changed soil hydrology—changes have been made that affect water flow in or out of the site.

Root cuts—the root system has been cut or otherwise significantly damaged. Additional information on root cuts will be included in the Roots and Root Collar box.

Describe—note applicable details or further descriptions of site changes.

Soil conditions—factors that can affect the ability of the root system to mechanically support the tree, as well as the general health and vitality of the tree. Check all that apply:

Limited volume—soil volume limited by rocks, water table, building foundations, size of a container, or other factors.

Saturated—soil saturated due to poor drainage, high water table, excess irrigation, or location in a low area. May be saturated now or have a history of inundation.

Shallow—rooting depth limited by one or more factors including high water table, rock ledges, compacted layers, or underground structures such as parking decks.

Compacted—soil is severely compacted, limiting the depth, spread, and distribution of the root system.

Pavement over roots—concrete, asphalt, pavers, or other materials restricting root growth or water movement into the root zone. If present, enter the percentage of the area within the drip line that is paved.

Describe—note applicable details or further descriptions of site conditions.

Prevailing wind direction—a typical, consistent, moderate-to-strong wind, usually from a single direction, that has affected tree crown and root system development.

Common weather—trees will adapt to a number of climatic conditions if they occur regularly. Check all that apply (strong winds, ice, snow, or heavy rain).

Describe—note any further descriptions regarding common weather.

Section 4—Tree Health and Species Profile

Tree Health and Species Profile
Vigor Low ☐ Normal ☐ High ☐   **Foliage** None (seasonal) ☐   None (dead) ☐   Normal _____ %   Chlorotic _____ %   Necrotic _____ %
Pests/Biotic _____   **Abiotic** _____
Species failure profile Branches ☐ Trunk ☐ Roots ☐ Describe _____

This section provides the opportunity to note any species-specific failure patterns that you suspect may influence likelihood of failure. Any species information you feel is important should be noted in this section. Any of these factors can be further described in the spaces provided or on additional paper.

Vigor—an assessment of overall tree health. Classify as low, normal, or high:

   Low—tree is weak, growing slowly, and/or under stress.

   Normal—tree has average vigor for its species and the site conditions.

   High—tree is growing well and appears to be free of significant health stress factors.

Foliage—size and color are indications of tree health; compare with a healthy specimen of the same species in the area:

   None (seasonal)—a deciduous tree that has dropped its leaves for the winter.

   None (dead)—a tree that has dropped its leaves because it is dead.

   Normal—percentage of foliage size and color that is normal for the species in the area.

   Chlorotic—percentage of foliage that is yellowish green to yellow.

   Necrotic—percentage of dead foliage in the crown.

Pests/Biotic—insects and diseases that may significantly affect tree health or stability.

Abiotic—abiotic problems that may significantly affect tree health or stability.

Species failure profile—any known failure problems with the species in the branches, trunk, or roots.

   Describe—note any further species failure details.

Section 5—Load Factors

Load Factors	
Wind exposure Protected ☐ Partial ☐ Full ☐ Wind funneling ☐ _____	**Relative crown size** Small ☐ Medium ☐ Large ☐
Crown density Sparse ☐ Normal ☐ Dense ☐ **Interior branches** Few ☐ Normal ☐ Dense ☐	**Vines/Mistletoe/Moss** ☐ _____
Recent or expected change in load factors _____	

Generally, two types of loads need to be considered when evaluating tree risk. Dynamic load is from wind as it impacts the tree, and static load is from gravity acting on the tree. These two loads can interact.

Wind exposure—factors that affect wind load on the tree. Check all that apply:

    Protected—trees or structures in the area significantly reduce wind velocity or the tree's exposure to wind.

    Partial—other trees, or buildings near the tree, moderately reduce the impact of wind on the tree.

    Full—tree is fully exposed to wind.

    Wind funneling—wind may be "funneled" or "tunneled" (by buildings, canyons, large stands of trees) toward the tree so that wind velocity experienced by the tree is increased.

Relative crown size—comparison of the tree's crown size to the trunk diameter. Classify as small, medium, or large.

Crown density—the relative wind transparency of the crown:

    Sparse—crown allows a large degree of wind and light penetration; varies with species.

    Normal—indicates moderate wind and light penetration.

    Dense—crown does not allow much light or wind penetration.

Interior branches—increase wind resistance but dampen branch/tree movement:

    Few—little wind resistance and damping.

    Normal—moderate wind resistance and damping.

    Dense—significant wind resistance and damping.

Vines/Mistletoe/Moss—check box if present at moderate to high levels that increase weight or wind resistance. Moss refers to Spanish or ball moss (epiphytes).

Recent or expected change in load factors—record any factors, recent or planned, that may significantly affect the load on any defects.

Section 6—Tree Defects and Conditions Affecting the Likelihood of Failure

```
— Crown and Branches —
Unbalanced crown ☐          LCR _____ %                    Cracks ☐ _____         Lightning damage ☐
Dead twigs/branches ☐    _____ % overall    Max. dia. _____    Codominant ☐ _____    Included bark ☐
Broken/Hangers    Number _____    Max. dia. _____    Weak attachments ☐ _____    Cavity/Nest hole ____ % circ.
Over-extended branches ☐                                   Previous branch failures ☐ _____   Similar branches present ☐
Pruning history                                            Dead/Missing bark ☐  Cankers/Galls/Burls ☐  Sapwood damage/decay ☐
  Crown cleaned  ☐        Thinned ☐         Raised ☐       Conks ☐           Heartwood decay ☐ _____
  Reduced        ☐        Topped ☐          Lion-tailed ☐  Response growth _____
  Flush cuts     ☐        Other _____
_____ Condition(s) of concern _____
_____
Part Size _____  Fall Distance _____    Part Size _____  Fall Distance _____
Load on defect    N/A ☐    Minor ☐  Moderate ☐  Significant ☐    Load on defect    N/A ☐    Minor ☐  Moderate ☐  Significant ☐
Likelihood of failure  Improbable ☐  Possible ☐  Probable ☐  Imminent ☐    Likelihood of failure  Improbable ☐  Possible ☐  Probable ☐  Imminent ☐
```

This section provides a systematic checklist for assessing the tree, dividing it into Crown and Branches, Trunk, and Roots and Root Collar. Check only factors that apply to the assessed tree. These factors may or may not contribute to Condition(s) of concern, Load on defect, or Likelihood of failure.

Crown and Branches

Unbalanced crown—check box if foliage is not uniformly distributed.

Live crown ratio (LCR)—the ratio of the height of the live crown to the height of the entire tree [LCR = (crown height/tree height) × 100].

Dead twigs/branches—small-diameter, dead branches. Check box if present and indicate percentage and size (maximum diameter).

Broken/Hangers—broken or cut branches remaining in the crown. Record the number and size (maximum diameter).

Over-extended branches—check box if there are branches that extend beyond the tree's canopy or that are excessively long with poor taper.

Pruning history—check appropriate boxes if pruning is known and relevant:

  Crown cleaned—pruning of dead, dying, diseased, and broken branches from the tree crown.

  Thinned—selective removal of live branches to reduce crown density. Other pruning types include, but are not limited to, structural, pollarding, espalier, and vista, and may be included in your notes.

  Raised—removal of lower branches to provide clearance.

  Reduced—pruning to decrease tree height or spread by cutting to lateral branches.

  Topped—inappropriate pruning technique used to reduce tree size; characterized by internodal cuts.

  Lion-tailed—inappropriate pruning practice removing an excessive number of inner and/or lower lateral branches.

  Flush cuts—pruning cuts through (or removal of) the branch collar, causing unnecessary injury to the trunk or parent branch.

  Other—note any other pruning history that may affect the likelihood of failure.

Cracks—separation in the wood in either a longitudinal (radial, in the plane of ray cells) or transverse (across the stem) direction. Check box if present and describe briefly.

Lightning damage—often evidenced by a centrally located line of sapwood damage and bark removal on either side in a spiral pattern on the trunk or branch. Check box if present.

Codominant—branches of nearly equal diameter arising from a common junction and lacking a normal branch union. Check box if present and describe.

Included bark—bark that becomes embedded in a union between branch and trunk, or between codominant stems, causing a weak structure. Check box if present.

Weak attachments—branches that are codominant or that have included bark or splits at or below the junctions. Check box if present and describe.

Cavity/Nest hole—openings from the outside into the heartwood area of the tree. Record the percentage of the branch circumference that has missing wood.

Previous branch failures—check box if there is evidence of previous branch failures and describe briefly. Check "similar branches present" if relevant.

Dead/Missing bark—check box if branches are dead or if areas of dead cambium are present where new wood will not be produced.

Cankers/Galls/Burls—check box if relevant and circle which one(s) are of concern:

> **Canker**—localized diseased areas on the branch; often sunken or discolored.
>
> **Gall**—abnormal swellings of tissue caused by pests; may or may not be a defect.
>
> **Burl**—outgrowth on the trunk, branch, or roots; not usually considered a defect.

Sapwood damage/decay—check box if there is mechanical or fungal damage in the sapwood that may weaken the branch, or decay of dead or dying branches. If checked, you may circle "damage" or "decay" to indicate which one is present.

Conks (mushrooms, brackets)—fungal fruiting structures; common, definite indicators of decay. Check box if present and describe under Condition(s) of concern.

Heartwood decay—check box if present and describe.

Response growth—reaction wood or additional wood grown to increase the structural strength of the branch. Note location and extent.

Condition(s) of concern—conditions in the crown and branches that may affect likelihood of failure. Note the main concern(s); if there are no concerns, write "none."

Part Size—a characterization of the part of the tree that may fail toward the target. Usually this is the diameter of the branch that can fall or the dbh of the tree. It may be appropriate to indicate the size of the part that could impact the target. Include units of measurement.

Fall Distance—if applicable, record the distance that the tree or tree part will fall before hitting a target; this may be relevant to the consequences of failure.

Load on defect—a consideration of how much loading is expected on the tree part of concern. Record as N/A (not applicable), minor, moderate, or significant, and/or note the cause of loading.

Likelihood of failure—the rating (*improbable, possible, probable,* or *imminent*) for the crown and branches of greatest concern. If there is a main concern, this information should be transferred to the Risk Categorization chart.

— Trunk —

Dead/Missing bark ☐   Abnormal bark texture/color ☐
Codominant stems ☐   Included bark ☐   Cracks ☐
Sapwood damage/decay ☐   Cankers/Galls/Burls ☐   Sap ooze ☐
Lightning damage ☐   Heartwood decay ☐   Conks/Mushrooms ☐
Cavity/Nest hole _____ % circ.   Depth _____   Poor taper ☐
Lean _____°   Corrected? _____
Response growth _____
Condition(s) of concern _____
Part Size _____   Fall Distance _____

Load on defect   N/A ☐   Minor ☐   Moderate ☐   Significant ☐
Likelihood of failure   Improbable ☐   Possible ☐   Probable ☐   Imminent ☐

Trunk

Dead/Missing bark—check box if a stem or codominant stem is dead or if areas of dead cambium are present where new wood will not be produced.

Abnormal bark texture/color—may indicate a fungal or structural problem with the trunk. Check box, if present, and add notes if it is a concern.

Codominant stems—stems of nearly equal diameter arising from a common junction and lacking a normal branch union. Note the size, location, and number, if relevant, under Condition(s) of concern.

Included bark—bark that becomes embedded in a union between branch and trunk, or between codominant stems, causing a weak structure. Check box if present.

Cracks—separation in the wood in either a longitudinal (radial, in the plane of ray cells) or transverse (across the stem) direction. Check box if present and describe under Condition(s) of concern.

Sapwood damage/decay—check box if there is mechanical or fungal damage in the sapwood that may weaken the trunk. If checked, you may circle "damage" or "decay" to indicate which one is present.

Cankers/Galls/Burls—may or may not affect the structural strength of the tree. Check box if present and circle which one(s):

    Canker—localized diseased area on the branch; often sunken or discolored.

    Gall—abnormal swelling of tissue caused by pests; may or may not be a defect.

    Burl—outgrowth on the trunk, branch, or roots; not usually considered a defect.

Sap ooze—oozing of liquid that may result from infections or infestations under the bark. May or may not affect structure or stability. Check box if present.

Lightning damage—often evidenced by a centrally located line of sapwood damage and bark removal on either side in a spiral pattern on the trunk or branch. Check box if present.

Heartwood decay—check box if present and identify/describe under Condition(s) of concern.

Conks/Mushrooms—fungal fruiting structures; common, definite indicators of decay when on the trunk. Check box if present and identify/describe under Condition(s) of concern.

Cavity/Nest hole—openings from the outside into the heartwood area of the tree. Record the percentage of the trunk circumference that has missing wood, and the depth of the cavity.

Poor taper—change in diameter over the length of the trunk, important for even distribution of mechanical stress. Check box if trunk has poor taper.

Lean—angle of the trunk measured from vertical. Record the degree of lean.

Corrected?—the tree may have been able to correct the lean with new growth in the younger portions of the tree. Note conditions related to lean in the space provided.

Response growth—reaction wood or additional wood grown to increase the structural strength of the trunk. Note location and extent.

Condition(s) of concern—conditions in the trunk that may affect likelihood of failure. Note the main concern(s); if there are no concerns, write "none."

Part Size—a characterization of the part of the tree that may fail toward the target. Usually this is the diameter of the branch that can fall or the dbh of the tree. It may be appropriate to indicate the size of the part that could impact the target. Include units of measurement.

Fall Distance—if applicable, record the distance that the tree or tree part will fall before hitting a target; this may be relevant to the consequences of failure.

Load on defect—a consideration of how much loading is expected on the tree part of concern. Record as N/A (not applicable), minor, moderate, or significant, and/or note the cause of loading.

Likelihood of failure—the rating (*improbable, possible, probable,* or *imminent*) for the trunk. If there is a main concern, this information should be transferred to the Risk Categorization chart.

— Roots and Root Collar —

Collar buried/Not visible ☐   Depth_____   Stem girdling ☐
Dead ☐   Decay ☐   Conks/Mushrooms ☐
Ooze ☐   Cavity ☐ _____% circ.
Cracks ☐   Cut/Damaged roots ☐   Distance from trunk _____
Root plate lifting ☐   Soil weakness ☐
Response growth _____
Condition(s) of concern _____

Part Size _____   Fall Distance _____
Load on defect   N/A ☐   Minor ☐   Moderate ☐   Significant ☐
Likelihood of failure   Improbable ☐   Possible ☐   Probable ☐   Imminent ☐

Roots and Root Collar

Collar buried/Not visible—check box if the root collar is not visible. If possible, determine and note the depth belowground.

Stem girdling—restriction or destruction of the trunk or buttress roots. Check box if it is a failure concern.

Dead—check box if one or more structural support roots are dead.

Decay—check box if present and identify/describe under Condition(s) of concern.

Conks/Mushrooms—fungal fruiting structures; common, definite indicators of decay. Fungal fruiting structures away from the trunk in the turf or mulch may be due to the presence of a mycorrhizal fungus and, if so, do not pose a threat to the tree. Check box if present and identify/describe under Condition(s) of concern.

Ooze—seeping or exudation that can result from pest infestations or infections under the bark. Check box if present and describe.

Cavity—definite indicators of heartwood decay. Measure the size of the opening and record the percentage of the tree's circumference affected.

Cracks—separation in the wood in either a longitudinal (radial, in the plane of ray cells) or transverse (across the stem) direction. Check box if present and describe.

Cut/Damaged roots—check box if present. Measure and record the distance from the trunk to the cut.

Root plate lifting—soil cracking or lifting indicates the tree has been rocking, usually in high winds. Check box if present, and note under Condition(s) of concern.

Soil weakness—check box if there is a soil condition affecting the anchorage of the tree's root system. Note under Condition(s) of concern if significant.

Response growth—reaction wood or additional wood grown to increase the structural strength of the roots or root collar. Note location and extent.

Condition(s) of concern—conditions in the trunk that may affect likelihood of failure. Note the main concern(s); if there are no concerns, write "none."

Part Size—a characterization of the part of the tree that may fail toward the target. Usually this is the diameter of the branch that can fall or the dbh of the tree. It may be appropriate to indicate the size of the part that could impact the target. Include units of measurement.

Fall Distance—if applicable, record the distance that the tree or tree part will fall before hitting a target; this may be relevant to the consequences of failure.

Load on defect—a consideration of how much loading is expected on the tree part of concern. Record as N/A (not applicable), minor, moderate, or significant, and/or note the cause of loading.

Likelihood of failure—the rating (*improbable, possible, probable,* or *imminent*) for the roots or root collar. If there is a main concern, this information should be transferred to the Risk Categorization chart.

PAGE 2—RISK CATEGORIZATION AND MITIGATION

The second page of the form focuses on categorizing the risk the tree poses and describing how the risk should be mitigated. It also provides space for additional notes or comments regarding any section from the first page. Use a separate sheet of paper if more space is needed.

Section 7—Risk Categorization

Risk Categorization

Target (Target number or description)	Tree part	Condition(s) of concern	Likelihood — Failure				Likelihood — Impact				Failure & Impact (from Matrix 1)				Consequences				Risk rating (from Matrix 2)
			Improbable	Possible	Probable	Imminent	Very low	Low	Medium	High	Unlikely	Somewhat likely	Likely	Very likely	Negligible	Minor	Significant	Severe	

Matrix 1. Likelihood matrix.

Likelihood of Failure	Likelihood of Impact			
	Very low	Low	Medium	High
Imminent	Unlikely	Somewhat likely	Likely	Very likely
Probable	Unlikely	Unlikely	Somewhat likely	Likely
Possible	Unlikely	Unlikely	Unlikely	Somewhat likely
Improbable	Unlikely	Unlikely	Unlikely	Unlikely

Matrix 2. Risk rating matrix.

Likelihood of Failure & Impact	Consequences of Failure			
	Negligible	Minor	Significant	Severe
Very likely	Low	Moderate	High	Extreme
Likely	Low	Moderate	High	High
Somewhat likely	Low	Low	Moderate	Moderate
Unlikely	Low	Low	Low	Low

This form uses the risk categorization methodologies presented in ISA's *Best Management Practices: Tree Risk Assessment*. The chart provided on the form is a tool to tie the data collected on the front of the form to the risk categorization process. You can rate the risk for up to four different conditions that may be found in the tree being assessed. Additional ratings may be made on an additional form. If there is only one condition of concern, only one line needs to be completed.

Target (Target number or description)—specify target number or a brief description from the first page of this form.

Tree part—specify the branch, trunk, or root of concern. For example, Condition Number 1 may be the broken branch over the house, and Condition Number 2 may be a branch over the driveway. The entries in the Tree part column would both be "branch." Other options for this column include "trunk" and "roots."

Condition(s) of concern—identify the concern(s) with the tree part listed. An example would be "large, dead branch over the house."

Tree risk has two components: (1) the likelihood of a tree failure striking a target, which is divided into the likelihood of failure and the likelihood of impact, and (2) the consequences of failure. Use your best judgment and the data available to assess the likelihood of failure (*improbable, possible, probable, imminent*) and the likelihood of impact (*very low, low, medium, high*). After these two decisions are made, use Matrix 1 (likelihood matrix) to determine the likelihood of failure and impact category (*unlikely, somewhat likely, likely, very likely*) based on your assessment.

The likelihood of failure can be categorized using the following guidelines:

Improbable—the tree or tree part is not likely to fail during normal weather conditions and may not fail in extreme weather conditions within the specified time frame.

Possible—failure may be expected in extreme weather conditions, but it is unlikely during normal weather conditions within the specified time frame.

Probable—failure may be expected under normal weather conditions within the specified time frame.

Imminent—failure has started or is most likely to occur in the near future, even if there is no significant wind or increased load. This is an infrequent occurrence for a risk assessor to encounter, and it may require immediate action to protect people from harm. The imminent category overrides the stated time frame.

Since these categories are time dependent, the time frame must be considered. The time frame is recorded on the first page.

The likelihood of impacting a target can be categorized using the following guidelines:

Very low—the chance of the failed tree or tree part impacting the specified target is remote. Likelihood of impact could be *very low* if the target is outside the anticipated target zone or if occupancy rates are rare. Another example of *very low* likelihood of impact is people in an occasionally used area with protection against being struck by the tree failure due to the presence of other trees or structures between the tree being assessed and the targets.

Low—there is a slight chance that the failed tree or tree part will impact the target. This is the case for people in an occasionally used area with no protection factors and no predictable direction of fall, a frequently used area that is partially protected, or a constant target that is well protected from the assessed tree. Examples are vehicles on an occasionally used service road next to the assessed tree, or a frequently used street that has a large tree providing protection between vehicles on the street and the assessed tree.

Medium—the failed tree or tree part could impact the target, but is not expected to do so. This is the case for people in a frequently used area when the direction of fall may or may not be toward the target. An example of a *medium* likelihood of impacting people could be passengers in a car traveling on an arterial street (frequent occupancy) next to the assessed tree with a large, dead branch over the street.

High—the failed tree or tree part is likely to impact the target. This is the case when there is a constant target with no protection factors, and the direction of fall is toward the target.

> Matrix 1 (likelihood matrix) is used to determine the combined likelihood of failure and impact in a given time frame. The resulting terms (*unlikely, somewhat likely, likely, very likely*) are defined by their use within the matrix and are used to represent this combination of occurrences in Matrix 2 (risk rating matrix).
>
> In the Consequences section, one category should be selected (*negligible, minor, significant, severe*). Consequences of failure are estimated based on the amount of harm or damage that will be done to a target. The consequences depend on the part size, fall characteristics, fall distance, and any factors that may protect the risk target from harm. The significance of target values—both monetary and otherwise—are subjective and relative to the client.

The consequences of failure can be categorized using the following guidelines:

Negligible—no personal injury, low-value property damage, or disruptions that can be replaced or repaired.

Minor—minor personal injury, low-to-moderate value property damage, or small disruption of activities.

Significant—substantial personal injury, moderate- to high-value property damage, or considerable disruption of activities.

Severe—serious personal injury or death, high-value property damage, or major disruption of important activities.

Risk rating—the risk rating of the individual part for a specified target. The risk rating is categorized using Matrix 2. Risk rating terms are *low, moderate, high,* and *extreme*.

Section 8—Notes, Mitigation, and Limitations

Notes, explanations, descriptions
_____
_____
_____
_____

Mitigation options
1. _____ Residual risk _____
2. _____ Residual risk _____
3. _____ Residual risk _____
4. _____ Residual risk _____

Overall tree risk rating   Low ☐   Moderate ☐   High ☐   Extreme ☐
Overall residual risk   None ☐   Low ☐   Moderate ☐   High ☐   Extreme ☐   **Recommended inspection interval** _____
Data ☐Final ☐Preliminary   **Advanced assessment needed** ☐No ☐Yes-Type/Reason _____
Inspection limitations ☐None ☐Visibility ☐Access ☐Vines ☐Root collar buried   Describe _____

Upon completion of the assessment, use this section to illustrate potential areas of concern and to offer mitigation options. Any further recommendations or notes should be included in this section.

Notes, explanations, descriptions—describe any conditions or factors that are not well described elsewhere on the form. Include notes on anything you need to take into consideration for making ratings or recommendations.

The grid, stem, and circle templates are provided for sketching any applicable details related to the tree or site.

Mitigation options—list options for mitigating each risk described. List your preferred recommendation on the first line.

Residual risk—the residual risk is for the risk remaining after the mitigation you are recommending. Residual risk can be *low*, *moderate*, *high*, or *extreme*.

Overall tree risk rating—the highest risk determined for the tree and target of concern. If there is more than one part or target rating, the tree risk rating is the highest of the group.

Overall residual risk—risk remaining if the highest-risk tree part is mitigated.

The shaded rows in the Risk Categorization chart may be used to assess residual risk after proposed mitigation. For each mitigation action, rate the expected risk remaining after treatment using the same methodology for categorizing risk as before.

Recommended inspection interval—recommended time for reinspection or inspection frequency.

Data—use these boxes to indicate whether this assessment is final or preliminary.

Advanced assessment needed—note the reason for any advanced assessment recommended.

Inspection limitations—factors that limited your ability to inspect the tree. Check all that apply and describe briefly.

Basic Tree Risk Assessment Form

Client _____ Date _____ Time _____
Address/Tree location _____ Tree no. _____ Sheet ____ of ____
Tree species _____ dbh _____ Height _____ Crown spread dia. _____
Assessor(s) _____ Tools used _____ Time frame _____

Target Assessment

Target number	Target description	Target protection	Target zone — Target within drip line	Target zone — Target within 1 x Ht.	Target zone — Target within 1.5 x Ht.	Occupancy rate 1–rare 2 – occasional 3 – frequent 4 – constant	Practical to move target?	Restriction practical?
1								
2								
3								
4								

Site Factors

History of failures _____ **Topography** Flat ☐ Slope ☐ _____% **Aspect** _____
Site changes None ☐ Grade change ☐ Site clearing ☐ Changed soil hydrology ☐ Root cuts ☐ Describe _____
Soil conditions Limited volume ☐ Saturated ☐ Shallow ☐ Compacted ☐ Pavement over roots ☐ _____% Describe _____
Prevailing wind direction _____ **Common weather** Strong winds ☐ Ice ☐ Snow ☐ Heavy rain ☐ Describe _____

Tree Health and Species Profile

Vigor Low ☐ Normal ☐ High ☐ **Foliage** None (seasonal) ☐ None (dead) ☐ Normal ____% Chlorotic ____% Necrotic ____%
Pests/Biotic _____ **Abiotic** _____
Species failure profile Branches ☐ Trunk ☐ Roots ☐ Describe _____

Load Factors

Wind exposure Protected ☐ Partial ☐ Full ☐ Wind funneling ☐ _____ **Relative crown size** Small ☐ Medium ☐ Large ☐
Crown density Sparse ☐ Normal ☐ Dense ☐ **Interior branches** Few ☐ Normal ☐ Dense ☐ **Vines/Mistletoe/Moss** ☐ _____
Recent or expected change in load factors _____

Tree Defects and Conditions Affecting the Likelihood of Failure

— Crown and Branches —

Unbalanced crown ☐ LCR _____%
Dead twigs/branches ☐ _____% overall Max. dia. _____
Broken/Hangers Number _____ Max. dia. _____
Over-extended branches ☐
Pruning history
Crown cleaned ☐ Thinned ☐ Raised ☐
Reduced ☐ Topped ☐ Lion-tailed ☐
Flush cuts ☐ Other _____

Cracks ☐ _____ Lightning damage ☐
Codominant ☐ _____ Included bark ☐
Weak attachments ☐ _____ Cavity/Nest hole ____% circ.
Previous branch failures ☐ _____ Similar branches present ☐
Dead/Missing bark ☐ Cankers/Galls/Burls ☐ Sapwood damage/decay ☐
Conks ☐ Heartwood decay ☐ _____
Response growth _____

_____ Condition(s) of concern _____

Part Size _____ Fall Distance _____ | Part Size _____ Fall Distance _____
Load on defect N/A ☐ Minor ☐ Moderate ☐ Significant ☐ | **Load on defect** N/A ☐ Minor ☐ Moderate ☐ Significant ☐
Likelihood of failure Improbable ☐ Possible ☐ Probable ☐ Imminent ☐ | **Likelihood of failure** Improbable ☐ Possible ☐ Probable ☐ Imminent ☐

— Trunk —

Dead/Missing bark ☐ Abnormal bark texture/color ☐
Codominant stems ☐ Included bark ☐ Cracks ☐
Sapwood damage/decay ☐ Cankers/Galls/Burls ☐ Sap ooze ☐
Lightning damage ☐ Heartwood decay ☐ Conks/Mushrooms ☐
Cavity/Nest hole _____% circ. Depth _____ Poor taper ☐
Lean _____° Corrected? _____
Response growth _____
Condition(s) of concern _____

Part Size _____ Fall Distance _____
Load on defect N/A ☐ Minor ☐ Moderate ☐ Significant ☐
Likelihood of failure Improbable ☐ Possible ☐ Probable ☐ Imminent ☐

— Roots and Root Collar —

Collar buried/Not visible ☐ Depth _____ Stem girdling ☐
Dead ☐ Decay ☐ Conks/Mushrooms ☐
Ooze ☐ Cavity ☐ _____% circ.
Cracks ☐ Cut/Damaged roots ☐ Distance from trunk _____
Root plate lifting ☐ Soil weakness ☐
Response growth _____
Condition(s) of concern _____

Part Size _____ Fall Distance _____
Load on defect N/A ☐ Minor ☐ Moderate ☐ Significant ☐
Likelihood of failure Improbable ☐ Possible ☐ Probable ☐ Imminent ☐

Risk Categorization

Target (Target number or description)	Tree part	Condition(s) of concern	Likelihood - Failure				Likelihood - Impact				Likelihood - Failure & Impact (from Matrix 1)				Consequences				Risk rating (from Matrix 2)
			Improbable	Possible	Probable	Imminent	Very low	Low	Medium	High	Unlikely	Somewhat	Likely	Very likely	Negligible	Minor	Significant	Severe	

Matrix 1. Likelihood matrix.

Likelihood of Failure	Likelihood of Impact			
	Very low	Low	Medium	High
Imminent	Unlikely	Somewhat likely	Likely	Very likely
Probable	Unlikely	Unlikely	Somewhat likely	Likely
Possible	Unlikely	Unlikely	Unlikely	Somewhat likely
Improbable	Unlikely	Unlikely	Unlikely	Unlikely

Matrix 2. Risk rating matrix.

Likelihood of Failure & Impact	Consequences of Failure			
	Negligible	Minor	Significant	Severe
Very likely	Low	Moderate	High	Extreme
Likely	Low	Moderate	High	High
Somewhat likely	Low	Low	Moderate	Moderate
Unlikely	Low	Low	Low	Low

Notes, explanations, descriptions
_____
_____
_____
_____
_____

North

Mitigation options
1._____ Residual risk _____
2._____ Residual risk _____
3._____ Residual risk _____
4._____ Residual risk _____

Overall tree risk rating   Low ☐   Moderate ☐   High ☐   Extreme ☐

Overall residual risk   None ☐   Low ☐   Moderate ☐   High ☐   Extreme ☐   Recommended inspection interval _____

Data ☐ Final ☐ Preliminary   **Advanced assessment needed** ☐ No ☐ Yes-Type/Reason _____

Inspection limitations ☐ None ☐ Visibility ☐ Access ☐ Vines ☐ Root collar buried Describe _____

This datasheet was produced by the International Society of Arboriculture (ISA) — 2017

Appendix 2
Common Wood Decay Fungi

The following table has been developed with input from plant pathologists around the world. Users must understand that the type and amount of decay, and the aggressiveness with which a decay fungus spreads, will vary considerably in different parts of the world. Variations also will occur with the species of tree affected and the location of the fungus on the tree. Some fungi attack live wood, some attack dead wood, and some attack both. Some occur only on hardwoods or conifers, and some occur on both. Some decay fungi will kill bark and cambium, and some are limited to specific parts of a tree. Note that the mere presence of a fungus does not always mean that the tree or tree part is unstable. Knowing the basic characteristics will help to determine where to look and what to expect. For example, we would not see *Armillaria* on scaffold branches or *Cerrena unicolor* killing roots.

The table shows the most common wood-decaying fungi, but it does not include all known decay fungi. It should not be seen as a definitive reference; it is intended as a starting point to better understand fungi of interest. Users are cautioned that the names of many fungi change over time as classification is refined, and a fungus species can have several common names.

Species	Sap rot	Heart rot	White rot	Brown rot	Root/Butt	Root rot	Trunk	Branches	Cankers	Notes
Amylostereum laevigatum			●						●	
Armillaria spp.			●			●				Spreads by root contact
Bjerkanders adusta	●		●				●	●		
Ceriporiopsia rivulosa		●			●					
Cerrena unicolor	●		●				●		●	
Chondrostereum purpureum	●						●	●		
Climacodon septentrionalis		●	●				●			
Daedaelopsis confragosa			●		●					
Daedalea quercina		●		●	●		●			
Echinodontium tinctorium		●			●		●			
Fistulina hepatica		●		●			●			
Fomes fomentarius	●	●	●				●			
Fomitopsis pinicola		●		●	●		●			
Fomitopsis officianalis		●		●			●			
Ganoderma adspersum		●	●				●			
Ganoderma applanatum		●	●		●		●			Spreads by root contact
Ganoderma australe			●				●			
Ganoderma lucidum			●		●					
Ganoderma resinaceum			●		●					
Globifomes graveolens			●							Hardwoods only
Grifola frondosa			●		●					
Hericium abietis		●			●					
Hericium erinaceus			●				●			

Species	Sap rot	Heart rot	White rot	Brown rot	Root/Butt	Root rot	Trunk	Branches	Cankers	Notes
Heterobasidion annosum			•		•					Spreads by root contact
Inonotus andersonii	•	•	•				•			
Inonotus dryadeus			•		•					
Inonotus dryophilus	•	•	•				•			
Inonotus hispidus	•	•	•				•		•	
Inonotus obliquus									•	Spreads by root contact
Inonotus tomentosus			•		•		•			
Irpex lacteus			•					•		
Kretzschmaria deusta			•		•		•		•	Formerly *Ustulina deusta*
Laetiporus gilbertsonii		•		•	•		•			
Laetiporus sulphureus		•		•			•			
Lenzites betulina			•							
Meripilus giganteus			•		•					
Oxyporus latemarginatus			•			•				Mainly hardwoods
Oxyporus populinus		•	•				•			
Perenniporia fraxinophilia			•				•			
Perenniporia subacida		•			•					
Phaeolus schweinitzii				•	•		•			
Phellinus everhartii		•	•				•			
Phellinus gilvus		•	•				•			
Phellinus hartigii		•	•				•			
Phellinus igniarius		•	•				•			
Phellinus noxius				•	•					
Phellinus pini		•	•				•			
Phellinus robineae			•				•			
Phellinus robustus		•	•		•		•			
Phellinus spiculosa		•	•						•	
Phellinus tremulae		•	•				•			
Phellinus weirii					•					Spreads by root contact
Pholiota squarrosa			•		•		•			
Piptoporus betulinus		•		•			•			Common on dead trees but can be cause of death
Pleurotus ostreatus			•				•			
Polyporus squamosus		•	•				•			
Postia sericeomollis		•		•	•		•			
Rigidoporus ulmaris				•			•	•		
Schizophyllum commune	•		•				•		•	
Sparissus crispa				•	•					
Spongipellis delectans		•		•			•			
Stereum sanguinolentum			•				•			
Thielaviopsis paradoxa			•				•			Palm trees

Glossary

acceptable risk—the degree or amount of risk that the owner, manager, or controlling authority is willing to accept.

acceptable risk threshold—the highest level of risk that does not exceed the owner/manager's tolerance.

advanced assessment (Level 3)—an assessment performed to provide detailed information about specific tree parts, defects, targets, or site conditions. Specialized equipment, data collection and analysis, and/or expertise are usually required.

adventitious branch—branch arising from a stem or parent branch and having no connection to apical meristems.

adventitious root—roots arising from roots or stems and having no connection to apical meristems.

aerial inspection—inspection of the upper tree parts not readily accessed from the ground; typically done by climbing or from an aerial lift.

aerial patrol—overflights of a utility right-of-way, large areas, or individual trees in a defined area to record the location of trees that are likely to fail and cause harm.

analysis—detailed examination of the elements or structure of something.

angiosperm—plant with seeds borne in an ovary. Consists of two large groups: monocotyledons (grasses, palms, and related plants) and dicotyledons (most woody trees, shrubs, herbaceous plants, and related plants) (contrast with *gymnosperm*).

annual rings—rings of xylem that are visible in a cross section of the stem, branches, and roots of some trees. In temperate zones, the rings typically represent one year of growth.

barrier zone—a chemical and anatomical barrier formed by the cambium present at the time of wounding in response to wounding. Inhibits the spread of decay into xylem tissue formed after the time of wounding. Wall 4 in the CODIT model (contrast with *reaction zone*).

basal swelling—increased wood growth in the area near or where the trunk and roots come together.

basic assessment (Level 2)—detailed visual inspection of a tree and surrounding site that may include the use of simple tools. It requires that a tree risk assessor inspect completely around the tree trunk looking at the visible aboveground roots, trunk, branches, and site.

bending moment—a turning, bending or twisting force exerted by a lever, defined as the force (acting perpendicular to the lever) multiplied by the length of the lever (see *moment*).

biomechanics (tree biomechanics)—the study of the action of forces on living trees.

bow—leans characterized by the top of the tree bending over more than the lower trunk, creating a curve.

brace rod (rigid brace)—metal rod used to support weak sections or crotches of a tree.

bracket—the fruiting body of a decay fungus (see *conk*).

breach of duty (of care)—failure to take reasonable care to avoid injury or damage to a person or property in a situation where the law imposes a duty of care.

brown rot—fungal wood rot characterized by the breakdown of cellulose (contrast with *soft rot* and *white rot*).

buckling—failure mode characteristic of collapsing under compressive stress.

bulge—swellings on branches, trunks, or root flares; often caused by new tissue formed as a response to movement and that reinforces the wood structure at the weak area.

buttress root—roots at the trunk base that help support the tree and equalize mechanical stress.

butt rot—decay of the lower trunk, trunk flare, or buttress roots.

cabling (flexible bracing)—installation of steel or synthetic cable in a tree to provide supplemental support to weak branches or branch unions.

callus—undifferentiated tissue formed by the cambium, usually as the result of wounding (contrast with *woundwood*).

cambium—thin layer(s) of meristematic cells that give rise (outward) to the phloem and (inward) to the xylem, increasing stem and root diameter.

canker—localized diseased area on stems, roots, and branches. Often shrunken and discolored.

cavity—open or closed hollow within a tree stem, branch, or root, usually associated with decay.

cellulose—complex carbohydrate found in the cellular walls of the majority of plants, algae, and certain fungi.

center of force—the point toward or from which a force acts; the resultant point of aggregate forces.

client—the person or organization for whom professional services are rendered. Usually, the owner or manager responsible for the trees.

clinometer—instrument used for measuring the height of a tree or other structure.

CODIT—acronym for Compartmentalization of Decay in Trees (see *compartmentalization*).

codominant (trees in a stand)—more than one tree sharing dominance in a stand of trees.

codominant stem—forked branches nearly the same size in diameter, arising from a common junction and lacking a normal branch union.

column of decay—wood decay inside a tree that extends longitudinally up and down a stem or through a branch.

compartmentalization—natural defense process in trees by which chemical and physical boundaries are created that act to limit the spread of disease and decay organisms (see *CODIT*).

compression—in mechanics, the action of forces to squeeze, crush, or push together any material or substance (contrast with *tension*).

compression crack—fracture caused by compressive stress.

compression wood—reaction wood in gymnosperms, and some angiosperms, that develops on the underside of branches or leaning trunks and is important in load bearing (contrast with *tension wood*).

conclusions—the summary and results of a risk assessment.

conk—fruiting body or nonfruiting body (sterile conk) of a fungus. Often associated with decay (see *bracket*).

consequences—outcome of an event.

consequences of failure—personal injury, property damage, or disruption of activities due to the failure of a tree or tree part.

constant occupancy—a target is present at nearly all times, 24 hours a day, 7 days a week.

corrected lean—tree lean characterized by a leaning lower trunk and a top that is more upright as a result of self-correction; sweep.

crack—separation in wood fibers; narrow breaks or fissures in stems or branches. If severe, may result in tree or branch failure.

decay—process of degradation by microorganisms.

decay–detection device—an instrument or tool developed to detect decay in tree parts.

decomposition—the breakdown or separation of a substance into simpler substances.

defect—an imperfection, weakness, or lack of something necessary. In trees, defects are injuries, growth patterns, decay, or other conditions that reduce the tree's structural strength.

definite indicator—an indicator that decay is definitely present.

degree of harm—the amount or extent of injury, damage, or disruption.

diameter—the length of a straight line through the center of a circle.

diameter tape—a measuring tape scaled such that when it encircles a tree trunk, the diameter can be read directly.

discoloration—wood response of a tree to microorganisms, including bacteria and non-decay-causing fungi, resulting in dead, darkened wood with little strength.

disruption—a delay or interruption of progress or continuity.

dominant (trees in a stand)—the tree or trees in a stand that are typically larger in height (taller), diameter, and crown spread than all the adjacent trees.

drag—wind resistance.

drip line—imaginary line defined by the branch spread of a single plant or group of plants.

drive-by (assessment)—limited visual inspection from only one side of the tree, performed from a slow-moving vehicle. Also may be called a windshield assessment.

duty of care—legal obligation that requires an individual to apply reasonable actions when performing tasks that may potentially harm others.

dynamic—study of how objects move under the action of forces.

earlywood—portion of an annual ring (growth ring) that forms during spring, characterized by large-diameter cells and thin walls. Also called springwood (contrast with *latewood*).

edge tree—a tree on the edge of a stand, growing under conditions of light and exposure different from those prevailing within the stand.

ethics—the body of moral principles or values governing a group or individual's conduct.

event—occurrence of a particular set of circumstances. In tree risk assessment, a tree or tree part falling and impacting a target.

extreme (risk rating)—defined by its placement in the risk rating matrix (Matrix 2); failure is *imminent* with a *high* likelihood of impacting the target, and the consequences of the failure are *severe*.

failure (of tree or tree part)—breakage of stem, branch, or roots, or loss of mechanical support in the root system.

failure mode—location/manner in which failure could or has occurred; for example, stem failure, root failure, or soil failure.

fiber—elongated, tapering, thick-walled cell that provides strength to wood.

fissure—a long, narrow opening or split.

flexure wood—response growth triggered by the continued flexing of a tree stem or branch.

force—any action or influence causing an object to accelerate/decelerate. Calculated as mass multiplied by acceleration. Is a vector quantity.

forest stand—a group of trees in a wooded setting.

freeze–thaw crack—frost crack; vertical split in the wood of a tree, often near the base of the bole, caused by internal stresses and low temperatures.

frequent occupancy—the target zone is occupied for a large portion of the day or week.

frost crack—vertical split in the wood of a tree, often near the base of the bole, caused by internal stresses and low temperatures.

fungal fruiting structures—the reproductive structures of a fungus (conks, brackets, mushrooms).

fungus (pl. fungi)—group of organisms from the kingdom Fungi, including yeasts, molds, mushrooms, and smuts. Typically multicellular, saprophytic, or parasitic and lacking vascular tissue and chlorophyll. Reproduces vegetatively and by various types of spores borne in fruiting bodies.

girdling root—root that encircles all or part of the tree trunk or the tree's other roots, constricting the vascular tissue and inhibiting secondary growth and the movement of water and photosynthates.

grade changes—a topographic alteration to the surface of the ground.

gravity—the force that attracts a body toward the center of the earth.

ground-penetrating radar (GPR)—a nondestructive device that uses radar pulses to image the subsurface.

guy—1) a steel or synthetic-fiber cable between a tree or branch and an external anchor (another tree, the ground, or other fixed object) to provide supplemental support. 2) a steel cable between a utility pole and an external anchor (another pole, the ground, or other fixed object, which may sometimes be a tree) to keep the pole upright. Guys act in tension (contrast with *prop*).

gymnosperm—plants with exposed seeds, usually within cones (contrast with *angiosperm*).

hand pull test—a load test that involves installing a line in a tree, and then pulling and releasing the line several times to move the tree or branch. Most commonly used in a pre-work inspection, but can be used as part of an advanced tree risk assessment.

harm—personal injury or death, property damage, or disruption of activities.

hazard—situation or condition that is likely to lead to a loss, personal injury, property damage, or disruption of activities; a likely source of harm. In relation to trees, a hazard is the tree part(s) identified as a likely source of harm.

hazard tree (synonym, hazardous tree)—a tree identified as a likely source of harm.

heartwood—wood that is altered (inward) from sapwood and provides chemical defense against decay-causing organisms and continues to provide structural strength to the trunk. Trees may or may not have heartwood (contrast with *sapwood*).

heartwood rot—any of several types of fungal decay of tree heartwood, often beginning with infected wounds in the living portions of wood tissue. Also called heart rot.

high (likelihood of impact)—the failed tree or tree part is likely to impact the target. This is the case when there is a constant target, with no protection factors, and the direction of fall is toward the target.

high (risk rating)—defined by its placement in the risk rating matrix (Matrix 2); consequences are *significant* and likelihood is *very likely* or *likely*, or consequences are *severe* and likelihood is *likely*.

hydrology—study of the properties, distribution, and effects of water on the Earth's surface, underground, and in the atmosphere.

imminent (likelihood of failure)—failure has started or is most likely to occur in the near future, even if there is no significant wind or increased load. The imminent category overrides the stated time frame.

impact—(*v.*) striking a target or causing a disruption that affects activities.

improbable (likelihood of failure)—the tree or tree part is not likely to fail during normal weather conditions and may not fail in extreme weather conditions within the specified time frame.

included bark—bark that becomes embedded in a crotch (union) between branch and trunk or between codominant stems. Causes a weak structure.

inspection—an organized and systematic examination.

inspection frequency—the number of inspections per given unit of time (for example, once every three years).

inspection interval—time between inspections.

interior tree—a tree within a stand of trees, protected from wind exposure.

land disturbances—disruptions to a terrestrial site, community, or ecosystem that alters the physical environment.

land-use history—actions, events, or changes that have taken place on a site.

latewood—portion of an annual ring (growth ring) that forms during summer, characterized by small-diameter cells with thick walls. Summer wood (contrast with *earlywood*).

lean—predominant angle of the trunk from vertical.

legal precedent—a principle or rule established by a prior court or other decision-making body.

level(s) of assessment—categorization of the breadth and depth of analysis used in an assessment.

lever arm—the distance between the applied force (or center of force) and the point where the object will bend or rotate.

liability—something for which one is responsible. Legal responsibility.

LiDAR (Light Detection and Ranging)—a remote sensing method that uses laser technology to measure tree size and location in relation to the target of concern.

lightning protection system—hardware installed in a tree to conduct the charge of a lightning strike to ground.

lignin—organic substance that impregnates certain cell walls to thicken and strengthen the cell to reduce susceptibility to decay and pest damage.

likelihood—the chance of an event occurring. In the context of tree failures, the term may be used to specify: 1) the chance of a tree failure occurring; 2) the chance of impacting a specified target; and 3) the combination of the likelihood of a tree failing and the likelihood of impacting a specified target.

likelihood matrix—a tool for categorizing the combined likelihood of a failure impacting a target.

likelihood of failure—the chance of a tree or tree part failure occurring within the specified time frame.

likelihood of failure and impact—the chance of a tree failure occurring and impacting a target within the specified time frame.

likelihood of impact—the chance of a tree failure impacting a target during the specified time frame.

likely (likelihood of failure and impact)—defined by its placement in the likelihood matrix (Matrix 1); *imminent* likelihood of failure and *medium* likelihood of impact, or *probable* likelihood of failure and *high* likelihood of impact.

limitations—restraints or factors that restrict the precision, applicability, or extent of something.

limited visual assessment (Level 1)—a visual assessment from a specified perspective such as a foot, vehicle, or aerial (airborne) patrol of an individual tree or a population of trees near specified targets to identify conditions or obvious defects of concern.

live crown ratio (LCR)—the ratio of crown length to total tree height.

load—1) a general term used to indicate the magnitude of a force, bending moment, torque, or pressure applied to a substance or material. 2) cargo; weight to be borne or conveyed.

load testing—in advanced tree risk assessment, pulling tests to measure or observe the amount of inclination and/or deformation to assess stability.

low (likelihood of impact)—there is a slight chance that the failed tree or tree part will impact the target.

low (risk rating)—defined by its placement in the risk rating matrix (Matrix 2); consequences are *negligible* and likelihood is *unlikely*, or consequences are *minor* and likelihood is *somewhat likely*.

mallet—a broad-headed hammer made of wood, plastic, or resin used for "sounding" a tree.

mass damping—a process by which the amplitude of oscillations is reduced; in trees, motion created by the forces of wind or rigging operations may be reduced through branch movement.

matrix—a rectangular array of rows and columns used to facilitate problem solving or decision making.

mechanical stress—a measure of the internal forces acting within a deformable body; force per unit area.

mechanics—study of forces and their effects on bodies at rest or in motion.

medium (likelihood of impact)—the failed tree or tree part could impact the target, but is not expected to do so.

meristematic tissue—undifferentiated tissue in which active cell division takes place. Found in the root tips, buds, cambium, cork cambium, and latent buds.

minor (consequences)—minor personal injury, low- to moderate-value property damage, or small disruption of activities.

mitigation—in tree risk management, the process for reducing risk.

mitigation options—alternatives for reducing risk.

mitigation priority—established hierarchy for mitigation of risks based on risk ratings, budget, resources, and policies.

mobile target—a target that is in motion or intermittently moving.

moderate (risk rating)—defined by its placement in the risk rating matrix (Matrix 2); consequences are *minor* and likelihood is *very likely* or *likely*, or likelihood is *somewhat likely* and consequences are *significant* or *severe*.

moment—a turning, bending, or twisting force exerted by a lever, defined as the force (acting perpendicular to the lever) multiplied by the length of the lever.

movable target—target that can be relocated.

multiple risks—the concept that any tree, part, or failure mode could represent more than one type of risk.

negligence—failure to use reasonable care, resulting in damage or injury to another.

negligible (consequences)—no personal injury, low-value property damage, or disruptions that can be replaced or repaired.

neutral plane—an imaginary plane where there is neither tension nor compression.

occasional occupancy—the target zone is occupied by people or other targets infrequently or irregularly.

occupancy rate—the amount of time targets are within a target zone.

oozing—seeping or exudation from a tree cavity or other opening.

open-grown—a tree that has grown with exposure to wind and other elements from all directions.

overextended branch—branch that extends outside the normal crown area.

owner/manager—the person or entity responsible for tree management or the controlling authority that regulates tree management.

parenchyma—thin-walled, living cells essential in photosynthesis, radial transport, energy storage, and production of protective compounds.

pathogen—causal agent of disease. Usually refers to microorganisms.

patterns of failure—common modes of tree failure within a tree species or failure of multiple trees in a contiguous area that share similar site histories or environmental conditions.

phloem—plant vascular tissue that transports photosynthates and growth regulators. Situated on the inside of the bark, just outside the cambium. Is bidirectional (transports up and down) (contrast with *xylem*).

possible (likelihood of failure)—failure may be expected in extreme weather conditions, but it is unlikely during normal weather conditions within the specified time frame.

potential indicator—an indicator that decay might be present.

precipitation—any form of water that falls to the Earth's surface, such as rain, snow, or sleet.

prioritizing targets—a process for ranking targets according to importance or value.

probability—the measure of the chance of occurrence expressed as a number between 0 and 1 (0–100%), where 0 is impossibility and 1 is absolute certainty. Often expressed as a percentage.

probable (likelihood of failure)—failure may be expected under normal weather conditions within the specified time frame.

probe—a stiff, small-diameter rod, stick, or wire that is inserted into a cavity or crack to estimate its size or depth.

prop—rigid brace, acting in compression, to support a tree, tree branch, or utility pole. Prop pole (contrast with *guy*).

protection factors—structures, trees, branches, or other factors that would prevent or reduce harm to targets in the event of a tree failure.

pruning—removing branches (or occasionally roots) from a tree or other plant, using approved practices, to achieve a specified objective.

pruning cycle—in utility and municipal arboriculture, the time scheduled between pruning events that is established as a guideline for providing reasonable clearance between trees and conductors.

qualitative tree risk assessment—a process using ratings of consequences and likelihood to determine risk significance levels (e.g., *extreme*, *high*, *medium*, or *low*) and to evaluate the level of risk against qualitative criteria.

quantitative tree risk assessment—a process to estimate numerical probability values for consequences and to calculate numeric values for risk.

radius—distance from the center to the perimeter of a circle. One half of diameter.

ram's horn—inward curling formation of woundwood resembling the horns of a ram.

rare occupancy—the target zone is not commonly used by people or other mobile/movable targets.

rays—parenchyma tissues that extend radially across the xylem and phloem of a tree and function in transport, storage, structural strength, and defense.

reaction wood—wood formed in leaning or crooked stems, or on upper or lower sides of branches, as a means of counteracting the effects of gravity (see *compression wood* and *tension wood*).

reaction zone—natural boundary formed chemically within a tree to separate damaged wood from existing healthy wood. Important in the process of compartmentalization (contrast with *barrier zone*).

recommendations—one or many alternatives that are promoted to achieve a desired outcome, based on professional judgment.

reporting (risk assessment reporting)—presenting the client with a summary statement describing in detail the results of an assessment.

residual risk—risk remaining after mitigation.

resistance-recording drill—a device consisting of a specialized micro-drill bit that drills into trees and graphs resistance to penetration; used to detect internal differences in the wood, such as decay.

response growth—new wood produced in response to loads to compensate for higher strain in outermost fibers; includes reaction wood (compression and tension), flexure wood, and woundwood.

retain and monitor—the recommendation to keep a tree and conduct follow-up assessments after a stated inspection interval.

retrenchment—natural process during which an overly mature tree reduces its crown and increases its girth to consolidate resources and increase longevity; the deliberate process of reducing tree height to mimic natural processes.

rhizomorph—a root-like aggregation of fungal hyphae.

rib—longitudinal bulge of response wood growth.

risk—the combination of the likelihood of an event and the severity of the potential consequences. In the context of trees, risk is the likelihood of a conflict or tree failure occurring and affecting a target, and the severity of the associated consequences.

risk aggregation—the consideration of risks in combination.

risk analysis—the systematic use of information to identify sources and to estimate the risk.

risk assessment—the process of risk identification, analysis, and evaluation.

risk categorization—the process of assigning risk and risk factors to categories based on severity or hierarchy.

risk evaluation—the process of comparing the assessed risk against given risk criteria to determine the significance of the risk.

risk management—the application of policies, procedures, and practices used to identify, evaluate, mitigate, monitor, and communicate risk.

risk perception—the subjective perceived level of risk from a situation or object, often differing from the actual level of risk.

risk rating—the level of risk combining the likelihood of a tree failing and impacting a specified target, and severity of the associated consequences.

risk rating matrix—a tool for ranking and displaying risks by assigning ratings for consequences and likelihood.

risk tolerance—degree of risk that is acceptable to the owner, manager, or controlling authority.

root collar excavation (RCX)—process of removing soil to expose and assess the root collar (root crown) of a tree.

root rot—decay located in the roots; root decay is usually developed from the bottom up, and crown symptoms may or may not be visible.

saprophyte—organism that lives on and may act to decay dead organic matter.

sapwood—outer wood (xylem) that is active in longitudinal transport of water and minerals.

sapwood rot—decay located in the sapwood. Bark and/or cambium may be damaged or dead. Signs of this classification of rot are usually numerous, but small, fruiting bodies along the bark's surface are common.

scope of work—the defined project objectives and requirements.

seam—lines formed where two edges of bark meet at a crack or wound.

secondary xylem—xylem produced to the interior of the vascular cambium during secondary growth.

severe (consequences)—serious personal injury or death, high-value property damage, or major disruption of important activities.

shear—1) *n.* in mechanics, the movement or failure of materials, especially laminar material such as wood, by sliding side by side. 2) *n.* a tool used to cut small-diameter plant material, including secateurs and snips, as well as long-bladed hand tools and power tools used to cut hedges. 3) *v.* to cut; often used to describe cutting foliage or stems to a single plane, as in a hedge.

shear plane crack—a crack at the neutral plane between tension and compression stresses.

shell wall—the remaining solid wood around a cavity or internal wood decay.

significant (consequences)—substantial personal injury, moderate- to high-value property damage, or considerable disruption of activities.

soft rot—decay of plant tissues characterized by the breakdown of tissues within the cell walls (contrast with *brown rot* and *white rot*).

soil compaction—compression of the soil, often as a result of vehicle or heavy-equipment traffic, that breaks down soil aggregates and reduces soil volume and total pore space, especially macropore space.

soil depth—the vertical extent of soil present below ground.

somewhat likely (likelihood of failure and impact)—defined by its placement in the likelihood matrix (Matrix 1); *imminent* likelihood of failure and *low* likelihood of impact, or *probable* likelihood of failure and *medium* likelihood of impact, or *possible* likelihood of failure and *high* likelihood of impact.

sonic assessment—a process of measuring wood density, or other mechanical properties, using an instrument that transmits, receives, and records the velocity of sound or electric waves through wood.

sounding—process of striking a tree with a mallet or other appropriate tool and listening for tones that indicate dead bark, a thin layer of wood outside a cavity, or cracks in wood.

standard of care—degree of care that a reasonable person should exercise in performing duty of care; a measurement used to assess whether an individual acted in a reasonable manner.

static pull test—in advanced tree risk assessment, a load test that measures outermost fiber strain in the stem or branches, and/or inclination at the root flare, in response to a controlled pull.

static target—target that cannot be easily relocated.

strain—the deformation resulting from a stress, measured as a change in specimen length per unit of total length.

stratifying targets—a process for classifying targets according to importance or value.

strength loss—degradation of the ability to withstand mechanical stress.

stress—1) in Plant Health Care, a factor that negatively affects the health of a plant; a factor that stimulates a response. 2) in mechanics, a force per unit area.

structural defect—feature, condition, or deformity of a tree that indicates a weak structure or instability that could contribute to tree failure.

structural support system—a device or mechanism providing supplemental support to individual branches and/or entire trees.

subdominant (trees in a stand)—understory trees in a stand or forest with growth somewhat restricted by larger nearby trees.

suberin—a waxy substance present in some cell walls.

sudden branch drop (SBD)—sudden, unanticipated failure of a tree branch with little or no discernible defect; often associated with long, horizontal branches and warm temperatures.

suppressed (tree in a stand)—understory trees in a stand or forest with growth severely restricted by competing nearby trees.

sweep—corrected tree lean characterized by a leaning lower trunk and a top that has grown back toward vertical.

taper—change in diameter over the length of trunks, branches, and roots.

target—people, property, or activities that could be injured, damaged, or disrupted by a tree failure.

target-based actions—risk mitigation actions aimed at reducing the likelihood of impact in the event of tree failure.

target management—acting to control the exposure of targets to risk.

target value—the monetary worth of something; the importance or preciousness of something.

target zone—the area where a tree or tree part is likely to land if it were to fail.

tension—in mechanics, the action of forces to stretch or pull apart any material or substance (contrast with *compression*).

tension wood—a form of reaction wood in broadleaved trees (hardwoods) that forms on the upper side of branches or the trunks of leaning trees (contrast with *compression wood*).

time frame—time period for which an assessment is defined.

timeline—time period for recommended mitigation.

tomography—use of multiple sensors placed around a trunk or limb to record sound or magnetic waves traveling through the wood, with measurements resulting in a picture of internal density characteristics. Typically used in arboriculture to measure the extent of decay in trees.

topography—the land and water features of an area, including changes in elevation.

torsion—the action of twisting or being twisted.

tracheid—elongated, tapering xylem cell adapted for the support and transport of water and elements.

tree architecture—the structural form and shape of a tree.

tree-based actions—risk mitigation actions aimed at reducing the likelihood of tree failure.

tree conflict—an interference between the needs of a tree and society or infrastructure.

tree growth regulator—chemical that can be applied to trees that slows terminal growth by reducing cell elongation.

tree population—a defined set, group, or collection of trees.

tree risk assessment—a systematic process used to identify, analyze, and evaluate tree risk.

tree risk evaluation—the process of comparing the assessed risk against given risk criteria to determine the significance of the risk.

tree risk management—the application of policies, procedures, and practices used to identify, evaluate, mitigate, monitor, and communicate tree risk.

unacceptable risk—a degree of risk that exceeds the tolerance of the owner, manager, or controlling authority.

unlikely (likelihood of failure and impact)—defined by its placement in the likelihood matrix (Matrix 1); *possible* or *probable* likelihood of failure and *low* likelihood of impact, or *possible* likelihood of failure and *medium* likelihood of impact, or *improbable* likelihood of failure with any likelihood of impact rating, or any likelihood of failure rating with *very low* likelihood of impact.

vascular cambium—lateral meristem from which secondary xylem and secondary phloem originate (see *cambium*).

verbal report—oral report; results of the risk assessment delivered to the client orally.

very likely (likelihood of failure and impact)—defined by its placement in the likelihood matrix (Matrix 1); *imminent* likelihood of failure and *high* likelihood of impact.

very low (likelihood of impact)—the chance of the failed tree or tree part impacting the specified target is remote.

vessel—end-to-end, tube-like, water-conducting cells in the xylem of angiosperms.

veteran tree—a tree which, because of its great age, size, or condition, is of exceptional cultural, landscape, or nature conservation value.

visual assessment—method of assessing the structural integrity of trees using external symptoms of mechanical stress (such as bulges, reactive growth, etc.) and defects (cracks, cavities, etc.).

walk-by (assessment)—a limited visual inspection, usually from one side of the tree, performed as the tree risk assessor walks by the tree(s).

white rot—fungal decay of wood in which both cellulose and lignin are broken down (contrast with *brown rot* and *soft rot*).

wildlife habitat—an environment suitable for sustaining one or more species of animals.

wind exposure—exposure to the forces of wind.

wind load—the force on a tree or structure resulting from the impact of wind.

wind velocity—the speed of wind.

windthrow—uprooting and overthrowing of a tree caused by wind.

wood decay—the process of wood degradation by microorganisms.

work order—a written document detailing the work to be completed and authorizing performance of contracted work.

woundwood—lignified, differentiated tissues produced on woody plants as a response to wounding.

written report—a document with text, images, and/or references, delivered in print or electronic form, containing the results of the risk assessment.

xylem—main water- and mineral-conducting tissue in trees and other plants. Provides structural support (contrast with phloem).

Selected References

American National Standards Institute. 2017. *American National Standard for Tree Care Operations: Tree, Shrub, and Other Woody Plant Management—Standard Practices (Tree Risk Assessment a. Tree Failure)* (A300 Part 9). Londonderry, New Hampshire: Tree Care Industry Association.

Ball, David J., and Laurence Ball-King. 2011. *Public Safety and Risk Assessment: Improving Decision Making.* New York: Earthscan.

Bond, Jerry. 2006. Foundations of tree risk analysis: Use of the t/R ratio to evaluate trunk failure potential. *Arborist News* 15(6):31–34.

Bond, Jerry. 2011a. Tree load: Concept. *Arborist News* 20(1):12–17.

Bond, Jerry. 2011b. Tree load: Basic field analysis. *Arborist News* 20(2):24–26.

Bond, Jerry. 2013. *Best Management Practices: Tree Inventories* (2nd ed.). Champaign, Illinois: International Society of Arboriculture.

Costello, Larry, Gary Watson, and E. Thomas Smiley. 2017. *Best Management Practices: Root Management.* Champaign, Illinois: International Society of Arboriculture.

Cullen, Scott. 2005. Trees and wind: A practical consideration of the drag equation velocity exponent for urban tree risk management. *Journal of Arboriculture* 31(3):101–113.

Davies, Caroline, Neville Fay, and Charles Mynors. 2000. *Veteran Trees: A Guide to Risk and Responsibility.* Peterborough, UK: English Nature.

Gilman, Edward F. 2012. *An Illustrated Guide to Pruning* (3rd ed.). Clifton Park, New York: Delmar.

Gilman, Edward F., and Sharon J. Lilly. 2008. *Best Management Practices: Tree Pruning* (2nd ed.). Champaign, Illinois: International Society of Arboriculture.

Harris, Richard W., James R. Clark, and Nelda P. Matheny. 2004. *Arboriculture: Integrated Management of Landscape Trees, Shrubs, and Vines* (4th ed.). Upper Saddle River, New Jersey: Prentice Hall.

International Organization for Standardization (ISO). 2009. *International Standard: Risk Management—Risk Assessment Techniques.* IEC/FDIS 31010.

Jepson, Jeff. 2009. *To Fell a Tree: A Complete Guide to Successful Tree Felling and Woodcutting Methods.* Longville, Minnesota: Beaver Tree Publishing.

Kane, Brian, Dennis Ryan, and David V. Bloniarz. 2001. Comparing formulae that assess strength loss due to decay in trees. *Journal of Arboriculture* 27(2):78–87.

Keefer, Christine A. (Ed.). 2004. *A Consultant's Guide to Writing Effective Reports.* Rockville, Maryland: American Society of Consulting Arborists.

Lilly, Sharon. 2010. *Arborists' Certification Study Guide* (3rd ed.). Champaign, Illinois: International Society of Arboriculture.

Lonsdale, David. 2010. *Principles of Tree Hazard Assessment and Management.* London: The Stationery Office.

Luley, Christopher J. 2005. *Wood Decay Fungi Common to Urban Living Trees in the Northeast and Central United States.* Palmyra and Naples, New York: Urban Forestry LCC.

Mattheck, Claus, Klaus Bethge, and Karlheinz Weber. 2015. *The Body Language of Trees: Encyclopedia of Visual Tree Assessment.* Karlsruhe, Germany: Forschungszentrum Karlsruhe GmbH.

Pokorny, Jill D. (Coord. Author). 2003. *Urban Tree Risk Management: A Community Guide to Program Design and Implementation.* Publication No. NA-TP-03-03, USDA Forest Service, Northeastern Area, State and Private Forestry, St. Paul, Minnesota.

Rinn, Frank. 2011. Basic aspects of mechanical stability of tree cross-sections. *Arborist News* 20(1):52–54.

Rinn, Frank. 2013. Shell-wall thickness and breaking safety of mature trees. *Western Arborist.* Fall 2013, pp. 14–18.

Rinn, Frank. 2014. How much crown pruning is needed for a specific wind-load reduction? *Western Arborist.* Spring 2014, pp. 10–13.

Roberts, John, Nick Jackson, and Mark Smith. 2006. *Tree Roots in the Urban Environment.* London: The Stationery Office.

Scharenbroch, Bryant, E. Thomas Smiley, and Wes Kocher. 2014. *Best Management Practices: Soil Management for Urban Trees.* Champaign, Illinois: International Society of Arboriculture.

Schwarze, Francis W.M.R. 2008. *Diagnosis and Prognosis of the Development of Wood Decay in Urban Trees.* Rowville, Victoria, Australia: ENSPEC Pty Ltd.

Schwarze, Francis W.M.R., Julia Engels, and Claus Mattheck. 2000. *Fungal Strategies of Wood Decay in Trees.* Berlin: Springer.

Smiley, E. Thomas, A. William Graham Jr., and Scott Cullen. 2015. *Best Management Practices: Tree Lightning Protection Systems* (3rd ed.). Champaign, Illinois: International Society of Arboriculture.

Smiley, E. Thomas, and Sharon Lilly. 2014. *Best Management Practices: Tree Support Systems: Cabling, Bracing, Guying, and Propping* (3rd ed.). Champaign, Illinois: International Society of Arboriculture.

Smiley, E. Thomas, Nelda Matheny, and Sharon Lilly. 2017. *Best Management Practices: Tree Risk Assessment* (2nd ed.). Champaign, Illinois: International Society of Arboriculture.

The National Tree Safety Group. 2011. *Common Sense Risk Management of Trees: Guidance on Trees and Public Safety in the UK for Owners, Managers and Advisers.* FCMS024. Edinburgh: Forestry Commission.

Urban, James. 2008. *Up by Roots: Healthy Soils and Trees in the Built Environment.* Champaign, Illinois: International Society of Arboriculture.

van Prooijen, Gerrit-Jan. 2010. *Stadsbomen Vademecum (Tree Inspection and Examination).* Arnhem, The Netherlands: IPC Groene Ruimte.

Index

Locators in **bold** indicate Figures. Locators in *italics* indicate Tables.

A

acceptable risk, 123, 134–**135**, 135, 149
access restriction, 147
adaptation *see* response growth
advanced assessment, 23–24, 32, *52*
adventitious branches, 104, **104**
adventitious roots, 109, 110
aerial inspection, 18–19, **19**, 24–25, **24**
aerial lifts, 25
aerodynamics, 95
angiosperms, 67, 70, 80
annual rings, **24**, 68, **68**, 69, 70
ants, 85, 86, **86**
Armillaria, 77, 82, **85**
aspect, 58–59
assessors, tree risk, *4*

B

bark
    atypical, 86, 103, 109, 111
    embedded, 27
    included, 105–106, **106**
    oozing, 86, 110, 111
barrier zone, 80–81, **81**
basal decay, 82–83
basal rot, **83**
basal swelling, 83, 109
basic assessment, 4–6, 20–23, *52*
Beaufort scale, 55
bending moment, 90, 91, **91**, 92
binoculars, 20, 25
biomechanics, 89–98
boring, increment, 24
boundaries, property, 9
bow, 118, **118**
braces, 142, 143, 147
brackets, 76, 85
branch attachments, 103–106, **103**, **104**, 112
branch failure, 102–103
brown rots, 77–78, **78**
bulges, 103, 107–108, **108**
butt rot, 83
buttress roots, 29, 70, 103, 109, **110**

C

cables, 142, 143, 146, 147
callus, 70, 113
cambium, **67**, 68–69, **68**, 72, 81, **81**, 113
cankers, 76, 92, 95, 107
case studies
    decay, 88
    decay assessment, 28
    limited visual assessment, 20
    mechanics, 96
    mitigation, 150
    reporting, 161
    risk categorization, 136
    site assessment, 62
    target assessment, 45
    tree biology, 72
    tree inspection, 117
    tree risk assessment, 13
cavities
    and cross-sectional strength, 93
    and decay, 85, 103, 111–112, **112**
    shell wall failure, 81
    sonic assessment, 27
    and wildlife, **86**
cellulose, 67, **67**, 70, 72, 77, *77*, 78, **78**, 87
center of force, 90
clients, *4*, 5, 8, 10, 11, 157
CODIT, 26–27, 79–82, **79**, **80**, 111
codominant branches, 104–106, **105**, 108
codominant trees, 57, **57**
column of decay, 81
compartmentalization, 26, **26**, **81**, **82**. *see also* CODIT
compasses, 22

compression, 67, 72, 91, 92, **92**
compression cracks, 115
compression wood, 70, **71**
conflicts, 8
conifers *see* gymnosperms
conks, 76, 85
consequences of failure, 6–8, 17, 43–44, **44**, 46, 128, 129–130, **131**
consequences of impact, *128*
constant occupancy, 40, **40**, 126
construction damage, 51
corrected lean, 118, **118**
cracks, 111–116, **113**
    compression, 115
    freeze–thaw/frost, 115, **115**
    ribs, 103, 108, **108**, 116, **116**
    seams, 86, 111, 116, **116**
    shear plane, 108, 114, **114**
    sonic assessment, 27
crowns
    dieback, 109
    live ratio, 119
    and load assessment, 95, 97
    pruning, 142
    raising, 143
    unbalanced, 119

D

data analysis, 122–137
decay
    assessment, 25–29, **26**
    and cross-sectional strength, 93–94, **93**
    as defect, 110–112
    detection tools, 27
    incipient, 26, **26**
    indicators, 85–86
    location, 82–84
    patterns, 25
    progression, 78–79
    root, 29–30, **30**
    strength loss from, 28–29
    testing, **24**
    tree biology, 75–88
deciduous trees, 82

decomposition, 79
decurrent trees, 97
defects, 103–110
degree of harm *see* harm
digging tools, **21**
discoloration, 79
dominant trees, 57, **57**
drag, 91
drainage *see* hydrology
drills and drilling, 24, 25–27, **25**, **26**
drive-by assessment, 18, **18**
drought, 53
duty of care, 11
dynamic force, 91

E-F

earlywood, **67**, 68
edge trees, 58
epicormic shoots, 104, **104**
equipment *see* tools and equipment
erosion control, 145
ethics, 11
excavation, 29, **29**, 53, 60
excurrent trees, 97
exposure *see* tree exposure
failure, 5, 6, 8, 51, 52, **53**, 62. *see also* likelihood of failure
failure threshold, 28
fertilizer, 145
fibers, 67, 69, **71**, 77
filling, 53, 61
"flag" trees, **95**
flexure wood, 70
flooding, 53
flow charts
    basic assessment, **22**
    limited visual assessment, **19**
force, 90–91, **91**
forest stands, 57–58, **58**
freeze–thaw cracks, 115
frequent occupancy, 40, **40**, 41, 42, 126, **127**, 129
frost cracks, 115, **115**
fungi

analysis, **24**, **32**
decay, 76–78
fruiting structures, 83, 85, 109, 111

G

Ganoderma, 77, **85**
girdling roots, 103, 109, 110, **110**
ground-penetrating radar (GPR), 30
grading, 60
gravity, 90
ground covers, 61
ground-penetrating radar (GPR), 30
growth regulators, 142, 146, 147
growth rings *see* annual rings
growth strategies, 69–73
guys, 142, 147
gymnosperms
    biology, 67, 68, 77
    decay, 25, 78, 80, 82, 84
    response growth, 70

H

hand pull test, **24**, 31, **31**
hardwoods, 70
harm, 6, 7, 11, 43, 124, 129
hazard, 6
heartwood, 76, **83**, **110**
heartwood rots, 84, **84**
height/diameter ratio, 94, 119
hemicellulose, 77, **77**, 78, **78**, 87
hydrology, 51, 53, 58, 145

I-J

included bark, 105–106, **106**
increment boring, 24
indicators, decay, 85–86
injuries, trunk, 107
inspections, **12**, 13, 100–101, 160
interior trees, 58
irrigation, **59**, 145

L

laboratory analysis, 32, **32**
Laetiporus, 78, **85**
land use, 51
latewood, **67**, 68
laws, 11–12
lean, 24, 30, **30**, 58, 117–118, **118**
legislation, 11
level (tool), **30**
lever arm, 90, 91, **91**
liability, 11
LiDAR (Light Detection and Ranging), 19
lightning protection, 144, **144**
lignin, 67, **67**, 69, 70, 72, 77, **77**, 78, **78**, 87
likelihood matrix, 123
likelihood of failure, 6, 7–8, 17, 62
    branches, 103–106, **104**, **105**
    classifying, 102
    cracks, 113–116
    from decay, 111
    response growth clues, 72–73, **73**
    and risk assessment, 101
    risk categorization, 123–124, **125**, *128*
    roots, 109–110, **109**
    trunk, 107–108, **109**
likelihood of impact, 17, 42, 126–127, **127**, *128*
limited visual assessment, 9, 17–19, 52
live crown ratio (LCR), 119
load, **90**
    analysis/testing, 24, 30–31
    assessing, 95, 97
    definition, 90
    measuring, 92–93

M

magnifying glass, 20
mallet, 21, **21**
managers, tree risk, *4*
mass damping, 95
measuring tools, 20
mechanical stress, 71
mechanics, tree, 89–98
mitigation, 10, 141–153, **141**
mitigation, risk, 39
mobile targets, 39–40, **39**, 41–42
moment, 90
monitoring, 147, 148, 149
monocotyledons, 69
movable targets, 39, **39**
mulches, 61, 145
multiple branches, 106, **106**
multiple risks, 134
mushrooms, 85

N-Q

neutral plane, 92
occasional occupancy, 40–41, **41**, 126, **127**
occupancy rates, 32, 40–42, **40**, 41, 126, **127**, 129
oozing, 86, 110, 111
open-grown trees, 57, 58
overextended branches, 106, **107**
parenchyma cells, 67, 80
pathogens, 76
people as targets, **38**, 39, 41–42
pests, 18
Phaeolus, 78
phloem, 67–69, **67**, 79
photography, 22, 159
power lines *see* utilities
precipitation, 56–57
prioritizing targets, 46–47, **46**, **47**
probability, 6
probes, 20, **21**, 25, **59**
professional responsibilities, *4*, 11–12
props, 142, 143, 146, 147
protection factors, 42, 44, 46, **46**
pruning, 142–143, 146, 147, 149
qualitative risk assessment, 6–7
quantitative risk assessment, 6

R

radiation, 24
radius, 93
ram's horn, 71
rare occupancy, 40, 41, **41**, 42, 126, **127**, 129
rays, 67, **67**, 80
root collar excavation (RCX), 29, **29**
reaction wood, 70
reaction zone, **71**, 80, **81**
reports, 10, 157–158
residual risk, 141, 148–149
resistance-recording drills, 24, 26–27, **26**, 28
response growth, 26, **26**, 70–73, **73**, **82**, **83**, 86, 101, 108, **109**
retain and monitor, 147, 148, 149
retrenchment, 69
rhizomorphs, 82
ribs, 103, 108, **108**, 116, **116**
ridges, 108, **109**
risk
    categorization, 122–130
    evaluation, 134–137
    vs. hazard, 6
    levels, 132, **133**
    mitigation, 10, 39, 141–153, **141**
    multiple, 134
    perception, 134
    rating, 10, 123, 128–129, 130–134
    residual, 141, 148–149
    types of, 8
risk assessors, *4*
risk management, 3–4
risk managers, *4*
risk targets *see* targets
root(s)
    adventitious, 109, 110
    buttress, 29, 70, 103, 109, **110**

collar, 29, **29**, 110
damage, 53
decay assessment, 29–31
defects, 109–110
development, 59–60
failure, 56–57
flare, 109
fused, 83
girdling, 103, 109, 110, **110**
rots, **76**, 82–83
shallow, **59**, 60
and site assessment, **60**
and site disturbance, 61

S

safety, 12–13, 71, 149–150, 151
saprophytes, 76
sapwood, 76, 84, **84**, **110**
scope of work, 8–10
seams, 86, 111, 116, **116**
shear, 92, **92**
shear plane cracks, 108, 114, **114**
shell wall, 25, 29, 93–94
site assessment, 24, 32, 50–63
    exposure, 57–58
    factors to consider, 51–57
    history of failures, 52
    site disturbance, 60–61, 63
    soil, 59–60
    topography, 58–59
    weather, 52–57
site modification, 142, 145, **145**
slopes, 58
soft rots, 78, **78**
soil
    compaction, 59, 60, 145
    depth, 59–60, *62*
    moisture, 59, 60, *62*, 63
    mounding, 109
    quality, 59, 60
    and root development, 59–60
    and site assessment, 51, 53, 59–60, **59**, **61**

testing, 60
unstable, 58
volume, 59
sonic assessment, 24, **26**, 27–28, **27**
sounding, 21, **21**, 25
standard of care, 11–12
static pull test, 24, 31, **31**
static targets, 38, 39, **39**, 40
stem girdling *see* girdling roots
storms, 18, 24, 32, 54
stratifying targets, 46–47, **46**, **47**
streamlining, 91, 94
strength, 25, 28–29, 92, 93–94, **93**
stress, 71, 72, 91–92, 94
structural stability, 69
structural support systems, 142, 143, **143**, 146, 147
structures *see* occupancy rates; static targets
subdominant trees, 57, **57**
suberin, 81
sudden branch drop, 53, 102
suppressed trees, 57, **57**
sweep, 118, **118**

T

taper, 94, 119
target value, 38, 43
target zone, 12, 37, **37**
target-based mitigation, 141, 142
targets *see also* consequences of failure; likelihood of failure; likelihood of impact; occupancy rates
    advanced assessment, 32
    defined, 5, 37–38, **38**
    management, 141
    stratifying and prioritizing, 46–47, **46**, **47**
    types of, 38–40
tension, 67, 91, **92**
tension stress, 72
tension wood, 70
termites, 85, 86, 109, 111
thinning, 142
time frame, 10, 124–125
timelines, 152–153, 160, 162
tomography, 24, **26**, 27–28, **27**

tools and equipment, 19–22, 24, 25–28, **26**, 30, **30**
topography, 51, **53**, 58–59, *62*
torsion, 92, **92**
tracheids, 67, 69, **71**, 80
traffic *see* mobile targets
tree architecture, 117–119
tree biology
   decay, 75–88
   mechanics, 89–98
   wood structure, 66–74
tree exposure, 24, 53, 57–58, **57**, **58**, 60, *62*
tree health, 24, 69, 73
tree inventories, 18
tree populations, 18–19, 23, **23**
tree removal, 142, 144, **144**, *146*, *147*
tree risk assessment, 5
   advanced, 23–24, 32, *52*
   aerial inspection, 24–25
   approaches, 6–8
   basic, 4–6, 20–23, *52*
   inspection, 100–120
   internal decay, 25–29
   limitations, 159–160
   limited visual, 9, 17–19, *52*
   and load, 95, 97
   process, 10–11
   professional responsibilities, 11–12
   risk management, 3–4
   safety, 12–13
   scope of work, 8–10
   single vs. population, 23
tree risk assessors, *4*
tree risk managers, *4*
tree-based mitigation, 142, **142**, *146–147*
trenching, 53
trunk, 103, 107–108, 112
trunk flare, 103, **110**, **111**

U-V

utilities, 12, 17, 38, 60, **61**
vascular cambium *see* cambium
vehicles *see* mobile targets
verbal reports, 158
vessels, 67, 80
veteran trees, 144
videos, 159, **159**
vigor, 73, **73**
visual assessment, 9, 17–19, 101, **101**

W-X

walk-by assessment, 18, **18**
watersprouts, 104, **104**
weather, 32, 42, 52–57, **56**, 95. *see also* storms; wind
weather stations, 54
white rots, 77, **77**
wildlife habitat, 148, **148**
wind, 51
   Beaufort scale, 55
   buffers, 145
   force, 90–91, **91**
   and likelihood of failure, *62*
   load, 24, 58
   and site assessment, 52–56, **53**, **54**, **57**
   tree response to, 94–95
windthrow, **57**
wood structure
   anatomy and physiology, 67–74, **67**, **68**
   growth strategies, 69–73
   health and vigor, 73
   response growth, 70–73
work orders, 157
wounds, **87**
woundwood, 70–71, **71**, **81**, 113
written reports, 158, **160**
xylem, 67–69, **84**
xylem, secondary *see* wood structure